AF695711

Digital Geography

Geospatial Technologies in the Social Studies Classroom

A volume in
International Social Studies Forum
Richard Diem and Jeff Passe, *Series Editors*

Digital Geography

Geospatial Technologies in the Social Studies Classroom

Edited by

Andrew J. Milson
University of North Texas
and
Marsha Alibrandi
Fairfield University

IAP

INFORMATION AGE PUBLISHING, INC.
Charlotte, NC • www.infoagepub.com

Library of Congress Cataloging-in-Publication Data

Digital geography : geospatial technologies in the social studies classroom / edited by Andrew J. Milson and and Marsha Alibrandi.
p. cm. – (International social studies forum : the series)
Includes bibliographical references.
ISBN 978-1-59311-672-9 (pbk.) – ISBN 978-1-59311-673-6 (hardcover)
1. Social sciences–Study and teaching–Computer network resources. 2. Social sciences–Study and teaching–United States–Computer network resources. 3. Geographic information systems. I. Milson, Andrew J. II. Alibrandi, Marsha.
LB1584.7.D54 2007
300.78'5–dc22

2007042238

Copyright © 2008 Information Age Publishing Inc.

All rights reserved. No part of this publication may be reproduced, stored in a retrieval system, or transmitted, in any form or by any means, electronic, mechanical, photocopying, microfilming, recording or otherwise, without written permission from the publisher.

Printed and bound by CPI Group (UK) Ltd, Croydon, CR0 4YY

CONTENTS

PART III

Explorations of Theory and Research on Geospatial Technologies in Education

PART IV

Teacher Preparation for Journeys with Geospatial Technologies

PREFACE

Marsha Alibrandi
Fairfield University

It is within relatively recent memory that the great potential for Geographic Information Systems (GIS) in K–12 education was first recognized. We have characterized this as a landscape that has been transforming over the past two decades, and Andrew Milson has organized the following chapters into sections describing that transforming landscape and our traverses upon and changes to it. A collection such as the contributions in this book capture and, to a degree, map the terrain we have traveled thus far. Therefore, this volume presents the collective efforts and commitment of our authors and salutes those who have pioneered GIS in K–12 settings, and those who are now preparing lessons, workshops, curricula, dissertations and conducting research in this newly-tilled field.

We are very excited to bring this collection of authors and their work together at what appears to be a watershed moment for geospatial technologies in social studies education. Our goal was to present a balance of research, theory and practice, and we are grateful to the authors for their ground-breaking work in each of these areas. It is exhilarating to look at this collection as a whole since it represents a spectrum of the 'International Geospatial Education Community' discussed by Joseph Kerski in Chapter 3. This volume is designed for the International Geospatial Education Community and we hope that it will serve as a sort of map and time line of our collective journey.

We have organized the chapters into sections whose span and spectrum indeed represent our collective journey. It has been a combination of pleasure and inspiration to read the chapter submissions and to see the totality as a body of work of which we can be pleased to have achieved in a mere

Digital Geography: Geospatial Technologies in the Social Studies Classroom, pages ix–xiii
Copyright © 2007 by Information Age Publishing
All rights of reproduction in any form reserved.

decade and a half. Below, I briefly summarize the chapters and their relationships, and encourage readers to spend the time to read all of our individual and collective work as we invite ever more participants in this journey.

In *Part I: The Landscape of Geospatial Technologies in Education*, we provide overviews from social studies, national and state standards, and a call to the international GIS community. In "A Social History of GIS in Education, 1985–2007," Marsha Alibrandi and Tom Baker build upon Baker's oft-cited dissertation timeline to include some of the context in which this journey has progressed. We provide foreground for the discussion using Thomas Friedman's 'world is flat' metaphor and ground our discussion in Diffusion of Innovations theory and the NCSS theme of Science, Technology and Society. Interviews with Charlie Fitzpatrick, Joseph Kerski and Anita Palmer, retrace the social history of where we've been and project a bit about where we're going. Their call for research is a major refrain echoed throughout the book.

Andy Milson and Jennifer Roberts further unfurl the map of the landscape of GST in "The Status of Geospatial Technologies in U.S. High School Geography Standards." Milson and Roberts undertook an extensive search of state standards to determine which states have established supportive infrastructure for implementing GIS and geospatial technologies (GST), and how these may be and must be revised and enhanced to support awareness, application and analysis with GST across the curriculum.

Who better than the inimitable Joseph Kerski to beckon us toward participation in the International Geospatial Education Community? Kerski is a recognized international leader of GIS in education in communities from North American Tribal colleges to elementary schools, to countless web-based resources for educators in the US, the UK, Central and South America and the Pacific. In his first chapter in this volume, Kerski chronicles the development of what he terms the "International Geospatial Education Community," and sounds the call for greater involvement in this community.

In *Part II: Social Studies Expeditions with Geospatial Technologies*, the research, development, and implementation of geospatial technology applications for social studies classrooms are presented. The exciting expeditions described here have been developed by explorers who have charted previously uncharted territories for others. In "Beyond Interactive Mapping: Bringing Data Analysis with GIS into the Social Studies Classroom," Northwestern University researchers and collaborators Danny Edelson, David Smith, and Matthew Brown take us behind the scenes and into student inquiry using GIS maps to learn about movement in U.S. history. Working in Chicago-area schools through iterations of their GEODE project, the authors describe their "on the ground" (aka "in the classroom") experience and the development of the *My World* GIS software application for education.

In "GIS for History: A GIS Learning Environment to Teach Historical Reasoning," Josh Radinsky of the University of Illinois at Chicago opens a gateway for Internet-based GIS in U.S. history. Radinsky's chapter opens a door to all U.S. history classrooms to begin their journey using GIS with an online interface designed for immediate, intuitive exploration by students and teachers with no prior GIS experience. Using historic census data from 1790 forward, *GIS for History* also incorporates maps from *History Matters*, the Library of Congress' American Memory and the Gilder-Lehrman collections.

In "The World at the Student's Fingertips: Internet-based GIS Education Opportunities," Joseph Kerski characteristically expands our horizons by describing the advancements in Internet-based GIS. These emerging Internet-based tools represent a new horizon for geospatial technologies in social studies education, and Kerski provides us with a focused guide for using these new tools to investigate "the whys of where."

In "Geospatial Online Learning Activities for Middle School Students" the team of geographic educators and researchers, Sven Fuhrmann, Alan M. MacEachren, Mark Gahegan & Roger Downs describe the development of the *MapStats for Kids* web portal and learning activities. This work represents a model for how developers can overcome the barriers of school technology infrastructures and student knowledge to bring geographic concepts and statistics to adolescents in a multimedia environment.

In "The Digital Globe: Using *GoogleEarth* for Virtual Fieldtrips of Coastal Environments," David R. Green & Joanna Mouatt take us to a special place in the U.K.—a place well worth visiting and preserving, and therefore worth the virtual field-tripping to see up close. These authors discuss how digital globe technologies, such as *GoogleEarth* can be combined with a variety of support tools, such as *GE Graph*, to conduct both 'virtual' and 'real' scientific field investigations. Their British geographic tradition is inspirational.

In *Part III: Explorations of Theory and Research on Geospatial Technologies in Education*, we see the emergence of theoretical frameworks upon which future research agendas may be developed—we hope collaboratively. In "Diffusion of Innovations Theory: Framing IT and GIS adoption," Shannon White introduces the value of locating GIS and GST in education within Rogers' Diffusion of Innovations theory. Following in Kevin Lynch's tradition and grounded in Paivio's Dual Coding theory, findings from White's (2005) dissertation are presented. Selected interview and representational data illustrate the value of understanding the meanings made by both teachers and students of the social, vocational, educational, school and community-based contexts and applications of GIS, particularly as they relate to GIS adoption by schools.

In "Going Flat: The Changing Dynamics and Integrative Nature of Geography in the Digital Classroom," Jay Gatrell & John Conant apply the five themes of geography as a theoretical construct for framing their discussion of analog (place) and digital (space) representations. Citing the National Academies of Science, Engineering and Medicine report *Rising above the Gathering Storm: Energizing and Employing America for a Brighter Future* (2005), the authors incorporate National Council for Economic Education goals and activities that are excellent additions to the analytical and curricular applications available for use in Social Studies classrooms.

In "Coming of Age: Research and Pedagogy on Geospatial Technologies within K–12 Social Studies Education," Aaron Doering, George Veletsianos and Cassandra Scharber recall the roots of GST and GIS as constructivist tools; that is, tools that were developed for inquiry. The authors explore the types of research that can be used to describe and capture constructivist learning and teaching and recommend design-based research methodology as widely applicable and appropriate for GST education research.

In "Instructional Geographic Information Science: A Multi-Disciplinary Framework for Geospatial Technologies in Education," Rick Bunch, Elisabeth Nelson, Robert Earl Lloyd, Michael Kane and Thomas Tricot articulate a model of multi-disciplinary research on geospatial technologies in education. Instructional GIS (InGIS) is presented as a model for expanding the conceptual framework underlying GIS and GST in education. The authors call for integrating traditions of geographic education, spatial cognition and educational psychology and relating research to advance our understanding of GST in education.

In *Part IV: Teacher Preparation for Journeys with Geospatial Technologies*, we explore the importance of preparing teachers for GST expeditions with their students. In "Spatial Thinking: The Key to Success in Using Geospatial Technologies in the Social Studies Classroom," Sarah and Bob Bednarz identify spatial thinking as a critical component for successful GST integration into teachers' practice. Bednarz and Bednarz locate two new "Rs"—representation and reasoning—as central to spatial thinking, and cite the findings of the 2006 NRC report *Learning to Think Spatially* that spatial thinking is indeed interdisciplinary. In their chapter, Bednarz and Bednarz discuss their National Science Foundation funded research program *Advancing Geospatial Skills in Science and Social Science* (AGSSS), and apply the theory of Technological-Pedagogical-Content Knowledge as a framework for interpreting their project outcomes and future developments in the field.

In "Examining the Teacher's Role When Teaching with Geographic Information System (GIS)," Eui-kyung Shin takes us inside her prize-winning research on the teacher's role in teaching GIS to fourth graders. Using discourse analysis to locate the situted decision-making by the teacher critical to her learners' success, Shin masterfully articulates the context of situated learning with GIS and makes recommendations for teacher education.

As our journeys continue, New Hampshire Geographic Alliance Coordinator and NCGE Vice President Kristin Alvarez goes "Diving In: Preparing Teachers to Implement Geospatial Technologies in the K–12 Classroom." Alvarez chronicles her initial full-course approach to GST for in-service and pre-service teachers. Incorporating desktop GIS, IGIS, and GPS activities with visits from local agency representatives, both 'actual' and virtual experiences were integrated for teacher preparation. As a model for a semester-long course now in progress, Alvarez shares her students' implementations of GIS once they've "dived in."

From a broadest view of the social, spatial, temporal, and technological landscapes, to a the more closely investigated, and therefore higher resolution perspectives gained from our collective journeys, with descriptions of useful theoretical and conceptual models, tools and avenues of access and instruction, to reflective discussions of courses traversed, we present a map reflecting where we've been, discussions of where we'd like to go, the tools we'd like to see, and some of the processes we can incorporate. We invite the reader to travel with us onto the unfurling *terra incognita* of the future landscapes, and hope that the work we've done to prepare for that future will be of use.

ACKNOWLEDGEMENTS

Andy Milson and Marsha Alibrandi wish to thank the International Social Studies Forum series editors, Richard Diem and Jeff Passe, for supporting a volume in their series on Geospatial Technologies (GST) in Social Studies Education. We also appreciate the gentle guidance and support of George Johnson of Information Age Publishing. We are extremely grateful to the authors of the chapters in this volume for their timely contributions and pioneering work in integrating GST into K–12, teacher education and on the Internet as resources for social studies classrooms.

Andy Milson wishes to thank co-editor Marsha Alibrandi for her excellent ideas and keen perspective on the field of GST in social studies. Her enthusiasm and tireless efforts to support social studies teachers are truly inspirational. Andy also thanks Lisa Hettler, doctoral student in Curriculum and Instruction at UNT, for her editorial and technical assistance with the text and images of the manuscript. Lisa has completed a crash course in APA and DPI!

Marsha Alibrandi thanks co-editor Andy Milson for his collaborative spirit and organizational skills in keeping the project on time and managed with grace and humor. She also thanks Barbaree Duke for sound advice and assistance; Charlie Fitzpatrick, Anita Palmer, Joseph Kerski and co-author Tom Baker for their insights in interviews.

The book is dedicated and testimony to the work of Charlie Fitzpatrick who, as recognized by his peers and the thousands of teachers who have had contact with him either directly or indirectly, has been a kindly driving force working to bring GIS to K12 and to the Internet. All of us thank you, Charlie.

Digital Geography: Geospatial Technologies in the Social Studies Classroom, page xv
Copyright © 2008 by Information Age Publishing
All rights of reproduction in any form reserved.

LIST OF TABLES, FIGURES, AND MAPS

Tables

Figures

Digital Geography: Geospatial Technologies in the Social Studies Classroom, pages xvii–xix
Copyright © 2008 by Information Age Publishing
All rights of reproduction in any form reserved.

Maps

PART I

THE LANDSCAPE OF GEOSPATIAL TECHNOLOGIES IN EDUCATION

CHAPTER 1

A SOCIAL HISTORY OF GIS IN EDUCATION, 1985–2007

Marsha Alibrandi
Fairfield University

Thomas Baker
Environmental Systems Research Institute

The twentieth century concludes in much the same way as it began—with the redrawing of maps. However, the new maps are not so much of geographic territory, but of landscapes depicting new and developing networks of finance, people and culture... The ultimate work of education is to learn to be a human being...But as we struggle for identities we must be able to transcend the notions of territory and engage new concepts of energy and place.

—Hartoonian, 1996: 6–8.

Since 1994, geographic educators and pioneering teachers and students have integrated Geographic Information Systems (GIS) into various curricular applications (Kerski, 2000; Baker, 2002; Alibrandi, 2003; NRC, 2005). With the advent of Internet-based online GIS (or IGIS) applications, a new age of GIS integration has been facilitated (Baker, 2005; White, 2005; Milson & Alibrandi, in press). Multiple platforms and possibilities now exist for social studies educators to consider. This chapter provides a social history and timeline of contextual innovations and developments surrounding GIS integration relevant to classroom instruction. We discuss the

Digital Geography: Geospatial Technologies in the Social Studies Classroom, pages 3–37
Copyright © 2008 by Information Age Publishing
All rights of reproduction in any form reserved.

overlooked stepchild of the National Council for the Social Studies (NCSS) ten themes: Science, Technology and Society (STS) as a model for both our research and for integrating *all* of the NCSS ten themes using GIS technology within a framework of Diffusion of Innovations theory (Rogers, 2003) in an increasingly "flat world" (Friedman, 2005).

AN OPPORTUNITY TO REFLECT ON SCIENCE, TECHNOLOGY AND SOCIETY

It is a rare opportunity to take a moment in the history of a technology—any technology—to reflect on its impact, consider its implications at the threshold of a new century, and to imagine its place in a curriculum that has largely been resistant to change. From O.L. Davis (1981) to Barr, Barth and Shermis (1977), to calls from Shirley Engle, Michael Hartoonian and other presidents of NCSS, to the members of the Geography Education National Implementation Project (GENIP) (Association of American Geographers [AAG], National Council for Geographic Education [NCGE], National Geographic Society [NGS], etc.) calling for increased geography across the social studies, the curriculum has remained relatively stable in terms of subject areas covered.

We take this opportunity in history and in social studies education to reflect, critique, and act on a technology that is now ubiquitous. Every weather program, news report, newspaper, tax map, on-board GPS and Mapquest inquiry of the past decade are now joined by digital globe programs such as GoogleEarth that 'mash up' information from local areas—hotels and hospitals to real estate and theme parks—as people wander across the new landscapes opened by the Internet.

With the proliferation of available—or ubiquitous—technologies such as auto and cell phone Global Positioning Systems (GPS) and new Internet-based Geographic Information Systems (IGIS), those involved in GIS in K–12 education for the past decade and a half see this moment in our technological and social history as a threshold. In 2001, Alibrandi (2001) predicted a 15-year uptake curve for schools integrating GIS technologies. From the 1992 watershed event cited by our interviewees to this publication, it has been fifteen years. That prediction was based on the slow uptake of simple computer applications such as word processing—available in schools in the 1980s, but not widely adopted until ten years later. With the added barriers of test-driven curriculum standards bearing down on most schools, only those free of such demands—private, charter and magnet schools—were able to lead the integration of GIS into school curricula (Audet & Ludwig, 2000). Because GIS application is not textbook or test-driven, the forces of those industries that impact curricular decision-making cannot be ignored. Yet this book is a message of hope for GIS and social studies. Its contributors describe successful GIS integrations, and have reflected and articulated theoretical and practical applications to collectively locate this technology within social studies across the K–16 spectrum.

The collection of voices in this book has operated since the mid-1990s at the edges of social studies to incorporate various new technologies into its curricula. Therefore the book's temporal positioning does signify a certain critical mass of experiences and technical innovations that have accelerated instructional technology integration in social studies education. These changes in the surrounding context, in combination with the younger generation's facility with new technologies, offer opportunities to construct new landscapes of learning that capture students' curiosity and inquiry. Therefore, many of this volume's authors will be found throughout this chapter's narrative and time line which is an attempt to apply a chronology to the diffusion of the GIS in education innovation.

DIFFUSION OF INNOVATIONS THEORY IN A "FLAT WORLD"

As an introduction to the eclectic collection of educators, theoretical perspectives and approaches presented in this volume, we borrow from Thomas Friedman's 'flat world' theory and from a theoretical framework familiar to most geographers and many involved in technology and innovation networks. We also propose framing GIS integration using Everett Roger's *Diffusion of Innovations* (2003), a theory originally posed to describe the expansion of techniques, ideas, tools and practices—even words and concepts—across human populations. The diffusion of innovations incorporates humanity's activities from stone tool-making to the mastery of fire by the dispersion of early hominids across the continents to Internet and satellite-based technologies and their adoption or non-adoption by populations. Some diffusion occurs over geographic space, and more recently diffusion occurs in cyberspace and uses satellite technologies. Shannon White (2005) explains Rogers' diffusion theory as it relates to the Internet and to GIS adoption in schools (see White, this volume). When an innovation such as GIS in education is viewed as a process along a continuum or curve of adoption over time, the factors that either accelerate or impede that process are contextual. Some of these contextual factors are shown in our time line to help locate this particular innovation's trajectory over the past 15 years.

Our positioning in being able to provide a social history before the story is forgotten is a unique opportunity as well. The key players and facilitating agencies, organizations and industries now intersect across new landscapes of learning on the Internet. Three of these key players were interviewed for this social history. These 'key players' were selected for their broad vision and international work with thousands of educators in the new landscapes of GIS in education. Charlie Fitzpatrick, now Director of the Schools and Libraries program at GIS software industry leader Environmental Systems Research Institute (ESRI), is personally known to thousands of educators from the National Geographic Society's Summer Institutes and its 50 state Geographic Alliances. Joseph Kerski, author of two chapters in this volume, has collaborated with each of the three key players in integral ways. Dr. Kerski's was the second Ph. D. dissertation addressing GIS in education, and as a geographer for the United States Geological Survey for some 18 years, designed, implemented and reported on hundreds of GIS workshops, presentations, short courses and institutes taught everywhere from elementary school classrooms to U.S. tribal colleges to international institutes with ESRI Business partner-trainers, *GISetc.* Anita Palmer, founder of *GISetc* with her husband Roger, is co-author of the highly praised *Mapping Our World: GIS Lessons for Educators* (Malone, Palmer, & Voigt, 2005). Palmer has taken the challenge to develop a business specifically focused on training teachers to apply GIS in the classroom. In doing so, *GISetc* has offered trainings in the U.S., Costa Rica, and New Zealand. If the diffusion of this innovation were to be mapped in a GIS, the three educators interviewed would appear with great frequency on that map.

Throughout this volume, we use geographic terms such as *intersection* and *landscape* to reflect the node-and-link (or "node and path," Lynch, 1960) connections used in connectionist theory to describe neural networks (Kosslyn, 1994). These neural networks have their origins in neurobiology, and are reflected upon landscapes, both natural and man-made. Human nature has evolved from hominids following water. The hydrologic of water creates patterns, pathways and networks of fluency similar to those followed by neurotransmitters throughout the brain and nervous system. These neural pathways are how we connect schemas, and the patterns humans create on the physical landscape also share those properties. If connectionist theory is applied to human learning of self-in-space (Golledge & Stimson, 1997), then it may also be applied to humans creating and using geospatial technologies.

It could be argued that humans, having an immense biological capacity for communication, have developed their most advanced technologies to be able to see and communicate ever further, faster and with ever greater spatial precision. These developments are of course not accidental, but intentional developments and extensions of that human need to communicate by new means, opening new vistas of satellite and aerial imagery. The impacts of these innovations on our understandings of the planet's systems seem finally to have arrived in our homes, on our desktop computers, in our consciousness, and into our policies and future technologies.

The impact of the Internet as a "flattening" agent cannot be underestimated. In fact, that design was intended by the Internet Society, whose mission statement is: "to promote open development, evolution, and use of the Internet for the benefit of all people throughout the world" (http://www.isoc/mission/). The Society, which maintains the Code of Conduct for the web, also lists various histories and timelines of the Internet. Today the number of Internet users is over one billion, and those from Asia have far surpassed the number from North America. Internet statistics are available from (http://www.internetworldstats.com/stats.htm). For an extensive time line of the Internet, see (http://www.zakon.org/robert/internet/timeline/). In fact, there are so many time lines available on the web that students should be routinely asked to find them, print them, cut them up into sections and use them for interactive classroom activities to support sequential and historical thinking.

In *The World is Flat*, Thomas Friedman (2005) cites what he sees as significant (STS) events and developments that have made the shift in "power, authority and governance":

> ...the opening of the Berlin Wall [in 1989], Netscape, "work flow," outsourcing, off-shoring, open-sourcing, in-sourcing, supply chaining, in-forming and "the steroids" [digital, mobile, personal and virtual] amplifying them all reinforced one another...They just needed time to start working together in a complementary, mutually enhancing fashion. That tipping point arrived sometime around the year 2000. The net result of this convergence was the creation of a global, Web-enabled playing field that allows for multiple forms of collaboration—the sharing of knowledge and work—in real time, without regard to geography, distance, or in the near future, even language. No, not everyone has access yet to this platform, this playing field, but it is open to more people in more places on more days in more ways than anything like it ever before in the history of the world. *This is what I mean when I say the world has been flattened.* It is the complementary convergence of the ten flatteners, creating this new global playing field for multiple forms of collaboration. (pp. 176–177)

Journalist Friedman casts many of the recent developments of technology and society in terms of the world becoming flat, or in other words, upon a more level playing field. We loosely frame these STS impacts on economic, cultural, social and environmental domains as well as the civic, governmental, and therefore individual and institutional impacts these changes are having and will have—in other words, across every one of the domains of the social studies.

WORKING ASSUMPTIONS

While some of the NCSS themes have been overfed by the 15 billion dollar textbook and testing industries, STS, which is by definition a place for critical reflection on technology and its impacts both in societies and on the planet, has been an undernourished area in our field. In regard to disciplines, rather than feed the false controversy of history-vs.-any other field within the social studies, we suggest a working assumption that social studies is a multi-discipline—it is plural after all—and that those many disciplines within it and from whence social studies educators have come, are all a part of that multi-discipline. The benefits of this working assumption are that a) we may all share in the social construction of the social studies; b) the multi-discipline need not remain static, but can adapt to changing understandings and facilitate multiple perspectives; c) we may examine new relationships, be they inter-disciplinary, or globally political, economic, social, cultural; and d) new landscapes of relationships facilitated by recent technologies will impact how we view the phenomena and events that have and will become history; and that the re-viewing of these events and phenomena are essential to the human experience on the planet. Finally, the degree to which these benefits are encompassed by STS, past present and future, is a perspective (be it implicit or explicit) from which many of the following chapters have been experienced, researched, and written.

Another assumption shared by this volume's authors is that the social studies landscape can become enriched, more textured, holistic and dynamic by addressing its hollows—those areas that have been thin in our preparation as social studies educators—as well as in our classroom activities and even our conceptualizations of how the world actually works.

One of the limits of linear historical thinking is that it doesn't encompass the whole—the so-called "big picture." The reason, of course, is that events occur not only in time, but in space, and expanding spatial understandings literally gives more *dimension* to simple historical understanding. Bednarz and Bednarz' chapter addresses the implications and application of *spatial thinking* as introduced in the National Research Council's report on the field (2006) in detail, but from this early point in the volume, let us use the familiar construct of the NCSS ten themes as an organizational scheme.

In Table 1.1, we offer a starting point for adding *the asking of geographic questions* to each of the ten NCSS themes. These questions may seem simple, yet without posing them, students may not draw the spatial and contextual conclusions that may seem obvious, but are not, and one might argue, are less and less visible to this generation who spend more time in virtual environments than has any previous generation.

TABLE 1.1
Asking Geographic Questions, GIS and Other Resources

NCSS Theme	Asking Geographic Questions	GIS Resources Projects
I. Culture	From where did these people, culture and language migrate? From which 'cultural hearth' did their foods, tools and travel radiate? How has the language and culture changed? Due to what influences? To what different areas have the language and culture spread? What of this culture's artistic, architectural, or social practices are seen globally?	*ArcLessons* http://gis2.esri.com/industries/education/arclessons/arclessons.cfm MLA Language Map (US): http://www.mla.org/map_single
II. Time, Continuity, and Change	How has the land use here changed over time? How have the climate and the ecology changed through history? How have the political alignments and boundaries changed? How has the ecology played a role on the economy of the region? Where are the region's trading partners? Competitors?	Library of Congress *Panoramic Map Collection*. http://memory.loc.gov/ammem/pmhtml/panhome.html *Sanborn Fire Insurance Maps:* http://www.lib.berkeley.edu/EART/snb-intr.html) Shin, this volume *GIS for History*: http://www.gisforhistory.org/ *ArcLessons* *Perry-Casteneda Map Collection:* http://www.lib.utexas.edu/maps/ *David Rumsey Map Collection:* www.davidrumsey.com
III. People, Places, and Environments	What factors are unique in the physical and cultural landscapes? What are the significant rivers, ecosystems, and why are their locations important? Is this a landlocked or well-connected place? How has that shaped its economy? How have people changed the land? How sustainable are the practices used here?	*ArcLessons, especially:* http://gis2.esri.com/industries/education/arclessons/search_results.cfm?id=305 http://rockyweb.cr/usgs.gov/outreach/activity.html Malone, Palmer & Voigt (2003). *Mapping Our World*, ESRI

TABLE 1.2
Asking Geographic Questions, GIS and Other Resources

NCSS Theme	Asking Geographic Questions	GIS Resources Projects
IV. Individual Development and Identity	How does a person (male/female) construct a sense of self here? What are the linguistic, religious, historic and contemporary influences in the region? Where did these influences come from? How and when did these influences come, and how is that changing?	Using "life maps," "Mental Maps" and "Windshield surveys" Alibrandi, 2003, Chapters 3–4
V. Individuals, Groups, and Institutions	Where did the influences of significant individuals come from? How did and do various groups interact? Are there conflicts over geographic resources? How are these resolved or unresolved? How do institutions (governments, religious groups, etc.) influence the political geography of the region?	*Mapstats for Kids* (http://wwww.fedstats.gov/kids/mapstats/ see Fuhrmann et al., this volume)
VI. Power, Authority, and Governance	Which polities govern which parts of civil life? (village, town, city, state, territory, province, nation or country? For example, how is schooling funded in your community? How would it compare with another system or state? What types of government have existed in this region's past and how do those relate to its current political structures?	See, Gattrell & Conant, this volume National Atlas: http://nationalatlas.gov/natlas/

VII. Production, Distribution, and Consumption	What products come from where? Are the production activities mainly primary, secondary, tertiary or quaternary activities? Who owns the means of production? Which products are produced multi-nationally? Who and where are the major trading partners? How environmentally sustainable is the production? What plans are in place for more energy efficient production?	*Globalis*: www.globalis.gvu.unu.edu National Atlas: http://nationalatlas.gov/natlas/ An Atlas of Cyberspaces http://www.cybergeography.org/atlas/
VIII. Science, Technology, and Society	From what traditions did the current technological practices evolve? What technologies are currently being developed? Are the current technologies sustainable? How do these impact the environment, urbanization, education, economics? Are there plans for transitions for more energy efficient production?	http://factfinder.census.gov >Maps> National Atlas http://nationalatlas.gov/natlasstrat.asp (see Map Layers: People; Economy; Energy Consumption; Transportation; Agriculture; Water, etc.)
IX. Global Connections	(all above) How do nations interact with one another? Alliances? What is a 'failed state?' What are the indicators of 'failure?' Why are many states considered 'multi-national?' How does the shape of a country affect its governance?	http:// www.globalis.gvu.unu.edu http://www.census,gov/ipc/www/idopyr.html http://webgis.wr.usgs.gov/globalgis http://education.usgs.gov/
X. Civic Ideals and Practices	From what cultural hearth have the ideals and traditions derived? What are the cultural practices considered civic ideals? How do different political units and structures interact? (VI above) How have these civic ideals and practices changed?	Perry-Casteneda Map Collection: http://www.lib.utexas.edu/maps/ *GIS for History*: http://www.gisforhistory.org/.

A NEW FOCUS ON SCIENCE, TECHNOLOGY, AND SOCIETY (STS), NCSS THEME VIII

With considerable foresight, NCSS itself, in the final decades of the 20th century, embraced as one of its ten themes, the idea of Science, Technology and Society (STS). Even ongoing debate within the scientific community itself over ethical violations and pressures, critical and feminist perspectives were finally acknowledged, and the sciences' positivistic tradition, in need of paradigmatic shift, began to inch toward more inclusive and flexible methodologies (Harding, 1986; Kwan, 2002).

As one of the ten social studies themes, Science, Technology and Society is perhaps less often taught than the more traditional standards such as Culture and Civic Ideals and Practices, or Power, Authority and Governance, but in this volume, we hope to demonstrate how Geographic Information Systems (GIS) and Internet-based GIS (IGIS) fully connect all of the ten themes—to one another and to other disciplines as well.

CONDUCTING SOCIAL STUDIES FOR THIS CHAPTER: CONSTRUCTING A TIME LINE AND SOCIAL HISTORY

The original structure of the following time line portion of this chapter is derived from Baker's (2002) dissertation, and has appeared in different versions in other publications. But we are adding new developments and more of the context of the enabling technologies to chronicle the sea change in which GIS technology is now found—on the sea of discourse that is the Internet. Baker (2005) has also written about the transition of personal computer-based GIS to Internet-based or *IGIS*, and that transition is not insignificant, nor is it over—we are in the midst of it as we write this in early 2007. Indeed, much of the research for this chapter and its expanded time line was researched on the web, where the short but eventful histories of these technologies reside. They are 'living' documents since the technologies are constantly in revision and are being in combined in ever new ways.

For the social history, ethnographer Alibrandi interviewed the movers and shakers of this educational innovation of integrating GIS. Those whose lives, energies, interests and talents are brought to certain courses of action are inextricable from its history. So have the social actors come, in the final months of 2006, to be collaborating as a newly expanded team at ESRI, the leading GIS software company that has supported the move to integrate GIS into schools. That team, originally led by Charlie Fitzpatrick, now includes George Dailey (previously with the Bureau of Census), Angela Lee, Dr. Joseph Kerski (formerly with the U.S. Geological Survey [USGS]) and co-author, Dr. Tom Baker who designed the first portal for GIS in education (http://www.KanGIS.edu) while at the University of Kansas. That Kerski and Baker chose to conduct dissertation research in GIS in education has had no small impact on the adoption of this innovation, and therefore, the move by ESRI to bring them into its leadership team signifies a continued commitment and garnering of talent for GIS advocacy. As of June 2007, the team had added a new dimension to ESRI's support of GIS in education, by creating the new http://edcommunity.esri.com with the mission of connecting and expanding the network of members of the GIS in education community.

Social History Interviews: Where Have We Been and Where Are We Going?

In conversations and interviews with 'movers and shakers' of the GIS in education movement, the most significant single event identified was ESRI's hiring of Charlie Fitzpatrick, who had been a central figure in the National Geographic Society's (NGS) $40 million, five year commitment to strengthen geographic education. Fitzpatrick had been the NGS institute's "technology person" for five years, helping teachers from across the U.S. to adopt the new technology tools available in the late 1980s for classroom use. An affable, patient and good-natured advocate for technology and geography, "Charlie" as he is known to everyone he meets, is a self-proclaimed geek who had taught 5th grade Geography in Minnesota.

While involved in the National Geographic Summer Institutes, he befriended the annual cadres of teachers who came in groups of about 100 each summer as the Society worked through the states, eight at a time for six weeks at a time over five summers in Washington, DC, where the participants shared classroom strategies, resources, field trips, worked on assignments in Charlie's computer lab, and studied for their final test. Charlie's wide network of geographic educators was already in place through National Geographic's state Geographic Alliances, still active since their inception (http://ngsednet.org/community/about.cfm?community_id=94). ESRI's hiring of a dedicated Schools and Libraries specialist signaled an important commitment from the GIS software company which was establishing itself as an industry leader. Other GIS software companies had also worked with specific teachers, but the combination of the existing NGS network and the introduction of GIS from a familiar and trusted colleague was a critical juncture. Subsequent to Charlie's position, two more GIS Schools and Libraries hires joined the team. Both George Dailey, formerly of the U.S. Census Bureau and a presenter at National Geographic's Summer Institutes, and Angie Lee were hired at that time so the demands of assisting teachers with *ArcView* and later products could be handled. Attending conferences, giving short courses, and assisting people by telephone and online were all needed in the early years, and this team provided that necessary assistance. For example, in "Impact and Success: Evaluation of the GIS Institute for Teachers," Sarah Bednarz (1999) reported that teachers felt that their questions were answered by Charlie, Joseph, and the other instructors both during the institute and subsequently in a listserv online.

Interview Questions

Because of the aforementioned and pre-existing familiarity and collaborations among most of members of the GIS in education community, the following interviews were conducted by Alibrandi as a participant observer from an *emic* or 'insider' perspective. Therefore the interviews were prearranged by email or telephone, conducted over email and/or telephone, based on the following questions:

1. From your perspective, what are the 3 most significant events or conditions that have facilitated integrating GIS into K–12 schools?
2. What do you consider the most successful initiatives for promoting GIS in K–12?
3. What do you think the GIS in K–12 agenda should be for 2007–2010?
4. How do you see GIS in K–12 in the future—say from 2010 to 2025?

There was clear crossover and consensus among the interviewees in the condition they signaled as most critical: teacher training. Each of the interviewees has conducted teacher training in a variety of formats from conference presentations to workshops, to institutes and to semester-long courses; in various softwares from IGIS to *ArcView 3.0* to *ArcGIS 9*; and in myriad locations from local to state to international audiences.

All three cited the initial ESRI-sponsored 'train-the trainer' institutes led by Charlie, George and Angie of ESRI with help from Joseph Kerski, veteran GIS teachers Kathryn Keranen and Reiny Freibertshauser and others in the summers of 1998, 1999 and 2000. These were mini-models of the National Geographic Summer Institutes for which Charlie had led the technology component. While Charlie cited the institutes as the most successful initiative (Question 2), both Joseph and Anita point to them as the most significant events (Question 1).

In interviews with Fitzpatrick, Joseph Kerski and Tom Baker, (now all working with ESRI as of December of 2006), the most significant events in promoting GIS in K–12 classrooms were those early institutes hosted in Texas in 1998 and 1999, and subsequently followed up by Kerski and interviewee Anita Palmer and her husband Roger who had been participants in the first institute. Others from that institute, Lyn Malone and Christine Voigt, joined Anita Palmer to write the very successful *Mapping Our World* GIS workbook for integrating GIS into World Geography classes with ESRI Press (2003, 2005). Fitzpatrick hails this book as one of the most significant additions to the field, and lauds the work of teammates Dailey and Lee as they designed the activities for what he fondly calls the GIS Boot Camps of 1998 and 1999. Indeed, the direct and sustained contact from those early institutes has had a lasting impact on the field, since Anita Palmer husband, Roger, along with Joseph Kerski and Tom Baker, have continued to offer workshops, institutes and online courses for GIS training through 2007.

Anita recalls the entrance requirements for the institute which were meant to establish a technology skill threshold for the applicants. "The initial application to the 98 institute—most of it was hard-core—can you produce a DBF file, can you zip a file? I mean, I had my students help me prepare that!"

Kerski recalls those initial GIS institutes as significant events:

> The 1998 and 1999 two-week GIS institutes that Charlie, George, Angie, and I conducted for educators at SW Texas (now TX State) University. Many of these educators are still the leaders in GIS-Ed and three went on to become authors of two ESRI Press books, which I would lump into this category, since the institutes spawned, eventually, the books. I would also lump into this category the many workshops that the Palmers and I conducted, and that Tom Baker and I conducted, before but especially after the institutes ended. The long 1–3 week summer institutes that the Palmers and I conducted from 2000 to 2006 were based on our desire that the 1999 institute would not be the last one—and because NGS nor anyone else was stepping up, that we would keep developing and hosting them.

The Palmers (Anita and Roger), subsequently developed the teacher trainings into their business, *GISetc* (http://www.gisetc.com/training.html), which has been an important force in the GIS in education field. Joseph mentions the two GIS curriculum books co-authored by Anita, Lyn Malone and Christine Voigt, the seminal *Mapping Our World: Geography Lessons for Educators* (aka: MOWGLE, 2002) and the *Community Geography Workbook* (2004). Fitzpatrick cites MOWGLE as the most significant event/condition from his perspective as a GIS software company representative working conferences and trade shows.

Charlie reflected, "Those people who became the 'Johnny Appleseeds' teaching other teachers in the field" were the goal of the original 1998–2000 institutes, starting their own waves of diffusion. Roger Palmer, originally from North Dakota, was instrumental in training teachers who became part of the group of Upper Midwest Aerospace Consortium (UMAC) states implementing NASA-based education programs into their curricula, committing to Internet in every classroom with four of the six states purchasing state-wide ESRI GIS site licenses for their schools. Cited by both Charlie and Anita, UMAC's five-state partnership is modeled somewhat after the original Geographic Alliances sponsored by National Geographic Society's (NGS) Geography Education initiative in the 1980s–90s.

Anita noted the important role of the still-functioning state Geographic Alliances as networks for teachers interested in integrating geography, and possibly GIS. Joseph cited that initiative and the current *My Wonderful World* initiative, led by National Geographic in its Partnerships for the 21st Century (see: http://www.mywonderfulworld.org/ and http://www.mywonderfulworld.org/educators_welcome.html), but noted the limited circle of influence of contributing organizations such as NGS, the National Council for Geographic Education (NCGE) and others. Regarding federal initiatives, Joseph noted:

> We're still feeling the initial effects, but I believe the Department of Labor statements about geotechnologies in 2004 and the subsequent President's Job Growth Initiative and the funding of GIWIS and Kidzonline (http://www.kidzonline.org/TechTraining/) and other projects designed to promote geospatial workforce development together with forces of globalization will have an impact on what gets supported, funded, and taught in the schools.
>
> Societal forces that can be summed up into a) environmental concerns brought to the attention of the general public in books such as *Last Child in the Woods*; and the high media attention on climate change; b) terrorism and the subsequent investment in geospatial data and intelligence, and the awareness of the fact that geotechnologies are important for understanding our world and living safely in it, and bound in that, a realization that students need to know this; c) the continual reports of Americans slipping in their lack of geographic understanding in the schools, coupled with globalization forces and the realization that we have an acute shortage of people who can think holistically and use technology effectively, particularly geotechnologies.

In addition, Anita reflected, "It takes many successful programs to move this forward. It's going to take more and more and more of these—one size does not fit all, and all of them make a difference—North Carolina, Florida you could point to every state and see that something has happened there, and someone has made a difference with each project." She cites several specific initiatives in many states and regions from Anne Olsen in New Zealand to the *AuthentiCITY* program in Phoenix, Arizona (see http://www.esri.com/news/arcuser/0700/auth.html); The GEODE Project that gave rise to *My World* GIS software (see Edelson, Smith, & Brown, this volume); Minnesota's *Eyes on Wildlife* (http://www.makingconnections.state .mn.us/eyes.html); the continuing role of Bob Coulter's Missouri Botanical Garden as a center for the National 4-H GIS/GPS Initiative (http://www.mobot.org/education/mapping/mapsp.html); American Forests, whose *CityGreen ArcView* and *ArcGIS* extension software was originally developed for planting trees for Atlanta's 1996 Summer Olympic Games (http://www.americanforests.org/productsandpubs/citygreen/school.php); and online initiatives such as the NSF-funded innovative online VISIT project and the recent Eastern Michigan University GIS Certificate program for teachers for which both Kerski and the Palmers were instructors. Continues Anita:

> We've trained a lot of people. The amount of people that actually come through and "sit down"—the multiplier effect—we're actually starting to see this. The first 2 years, we had almost nothing written down, but then we had *MOWGLE.* After that we saw a jump of 18–42% implementation after the training. They all left with an "Aha," and a 90% approval rating, but much less implementation. Last summer institute, I had a sense that 80–90% could use this—and we've heard from 505 teachers—"We're using this and we're training other teachers"—the seed has germinated, and it's very cool.
>
> Is that because we're better trainers? Better materials? They've heard of it—federal support, support from [school] administration, so there's a multiplier effect. We've got people calling us—"Will you hire us? Will you franchise?" So...we're getting lots of good feedback. We've resisted becoming authorized trainers for other than education, even though that could pay us better. We stay with this because we believe in it. We're the only two that are full time, and we can make a difference. So we're getting better at what does work—and there's of course as much to be learned by what doesn't work. We need research, that's for sure.

Imagining the GIS in education agenda for 2010 is a difficult question to answer when we consider the speed at which the Internet and other technologies have changed in the past two years. Digital Globe applications such as GoogleEarth, the new easy-to-use podcast and online video sharing technologies such as YouTube, and the new introduction of GPS-to-GIS tools from PASCO Scientific have all become widely available within about two years. Given the aforementioned seemingly exponential acceleration in growth and diffusion of the Internet, all bets might be off in terms of what to expect by 2010, but Fitzpatrick projects that the distribution of GIS data via Internet both has been and will be the most significant condition. Even with that distribution and Internet-based GIS (IGIS), he still believes:

> The challenge is identifying teachers and educational opportunities where exploration, inquiry and analysis are promoted, rather than step-by-step-by-step concrete lessons that will lead to a predetermined result. The problem that I see with a technology that is a "thinking technology" is that if it's approached as "tell me to how to use this" it's really antithetical... people trying to do plug-and-chug lessons—like using a Ferrari for commuter traffic.

Charlie recommends a sequence of learning to "explore maps and analyze with the data, and to graduate from something like *ArcWeb Explorer* to *AEJEE* to *ArcGIS*." Responding to this challenge, Anita adds:

> There's a debate about constructivism -vs-*MOWGLE*...but at least it got people off the block. Teachers have told us, "I love the materials, but now I need training"—so [*MOWGLE*]'s the training wheels, and then they can go on their own. We need curriculum development—across the content areas—working on lessons in math, science, Barbaree [Duke]'s Mark Twain activity—*we need curriculum development*—for both desktop and map machines. We think they both need curriculum across the content areas to make it more ubiquitous, so it becomes ingrained in every new initiative. Good published materials can make a big difference. People are looking for materials. We need to finish our book for elementary applications.
>
> One of the projects we're working on is an online Map book—Joseph, Tom, and Sarah Bednarz—looking at many of the wonderful online map machines. It's an idea book. What kinds of questions do your students ask? We think if it's anything, it's a way to get teachers to see and use online mapping. You can do a 15 minute session on an online map machine—it raises the level of wanting greater functionality. The tools can work in tandem to support each other, but with online mapping—what do you do with it? How many times can you find your house? The digital globes are very seductive, and could be nothing but a sexy game. But people play with it and say, "Oh, I can do this with this..." Online mapping has become part of the fabric, but no-one is really doing much with them in the classroom, so now we're looking at, 'How can we bring that in to develop spatial awareness?' But ask anyone and they can say, Oh, yeah, I'm doing GIS"—but now what do I do with this? We don't want people to stop. So what we need to continue to do is professional development, curriculum design, re-evaluate and development—changing with the times.

By 2001, Alibrandi, White, Duke, and Hagevik had introduced the concept of *constructivism* as a methodological framework for GIS in education at the ESRI EdUC based on the first two years of offering a semester-long GIS in Education course at North Carolina State University's College of Education. While some interpreted the use of *MOWGLE* as prescriptive rather than constructivist, the Alibrandi et al. view was and is that "all's fair in constructivism." In other words, when starting from the learner's prior knowledge, if he or she needs step-by step instruction, then *MOWGLE* is the perfect vehicle with its click-by-click instructions. If the learner is ready to apply GIS to a problem, then develop the appropriate skills.

Finally, Joseph Kerski weighs in on the agenda for the next few years in an email reply with his list of priorities:

1. More hands-on institutes.
2. Short web-casts and stored video help files that educators can access.
3. An educational portal for online community building.
4. An entire library of high quality, peer reviewed, Web-based GIS-geospatial thinking-web mapping courses offered at low cost for educators.
5. Books: GIS for teaching in business, math, history, environmental studies, earth science, and more.
6. Advocacy at all levels—district, state dept of education, Congress.
7. Partnerships with NGS, federal agencies, AAG, NASA, nonprofit, associations, industry, and many more organizations.
8. Outreach! Press releases, notices, news media, and more.
9. International collaboration with HERODOT, ICA, IGU, RGS, Geographical Association, UNEP, and more.

We think that this book helps to further those aims, and offer the following timelines to frame where we've been as we think about where we're going with GIS in education.

In the following timeline (see Figure 1.1), we have presented distinct tracks of activities that were related and simultaneous, though only at times coordinated across tracks. The purpose of the timeline is to chronicle the contextual events that surrounded early innovations in K–12 GIS. The timeline, taken as a whole is presented primarily as a reference.

In the annotated section following the timeline, Baker provides greater detail to some of the events, developments and curricular initiatives that comprise both the social history and the innovations within the technology itself.

Discussion and Annotated Timeline

As the use of Geographic Information Systems expanded throughout the 1980s, more and more discussion was directed at diverse industries, including K–12 education. In a 1987 paper, Thompson raised issues related to teaching *with* and *about* GIS in schools. He advocated careful consideration of teaching models and subject content that would be most advisable to use with GIS. While, Poiker (1985) recommended that certain attributes of GIS software should be identified as particularly valuable for school use, Poiker and Walsh suggested using "manual GIS" techniques, such as overlays, as precursor activities to implementing computer-based GIS.

	1980–1990	1990–1995	1996–2000	2001–2005	2006–2010
National Level Initiatives & Events	**1985:** The *Geographic Education National Implementation Project* (GENIP), a consortium of geographic associations committed to improving the status and quality of geography education in the United States launched (now at http://genip.tamu.edu/) **1986–1991:** (http://www .nsdc.org/library/publications/jsd/jsd11-97binko.cfm) *National Geographic Society*'s (NGS) Summer Institutes and *State Geographic Alliances* *Geographic Awareness Week* is launched	**1992:** Direct ESRI support for K–12 GIS. Charlie Fitzpatrick, formerly of the NGS Summer Institutes is hired by ESRI (Environmental Systems Research Institute), GIS software industry leader.[a] **1994:** *Geography for Life,* National Geography Standards Project **1994:** Global Learning and Observations to Benefit the Environment (GLOBE) program launched. Educational environmental data collection program using digital probes is made compatible with ESRI's *ArcVoyager* **1994:** ESRI hires George Dailey (formerly of US Census Bureau) and Angela Lee. **1995:** The *NatureMapping* Program whose vision is to create a national network linking natural resource agencies, academia and land planners with local	**1998 & 1999:** ESRI sponsors Authorized Teaching Program summer institutes at SW TX University. Many participants became leaders in GIS in Education, with 3 teaming as authors of *Mapping Our World* (ESRI Press, 2002).[b] **1999:** First *GIS Day,* national initiative by ESRI, NGS and Association of American Geographers (AAG) surpassed its goal of reaching 1 million as more than two million participated in events worldwide. **1997–2005:** KanCRN Kansas Collaborative Reseach Network for student research and collaboration online with immediate online mapping of student-entered data. (http://kancrn.org/index.cfm) **1999:** Roger Williams University (Hosted by Dr. Richard Audet) invited	**2001:** NSF funds what becomes the KanGIS web portal, designed by Tom Baker **2003:** 4H Adoption of GIS in its programs (see www.esri.com/4-H)	**2006:** *My Wonderful World* www.mywonderfulworld.org Multi-partner 5-year geographic education initiative hosted by National Geographic Society

		communities primarily through schools, wins the national RENEW America Award (http://depts.washington.edu/natmap/) **1995–1999:** Pennsylvania State/TERC project using NASA-sponsored EarthKam (see Barstow et al., 1999)	university faculty of education to a 1-week institute on GIS. C. Fitzpatrick introduced *ArcVoyager* and RI state and US GIS agents from the Census Bureau and other agencies described GIS integration initiatives. **1999:** First International Symposium on Digital Earth, held in China (see: http://www.isde5.org/)		
State of Technology	*TIGER* Census Data designed for database integration/sharing First *SPOT* satellite launched *GPS* becomes functional	**2/5/95:** *Mapquest* launches its www.mapquest.com website **8/9/95:** Netscape goes public See Friedman's *The World is Flat,* pp. 56–70 and for an historic document, see: www.nsf.gov/od/lpa/news/03/pr0343.htm **1997:** The Library of Congress releases the first of the Panoramic Maps Collection: (http://memory.loc.gov/ammem/pmhtml/panhome.html) as reported at: http://scout.wisc.edu/Reports/ScoutReport/1997/scout-970613.html)	Google founders, Larry Page and Sergey Brin as Stanford grad students begin collaborating on *BackRub*—a process for tracing back links from websites **2000:** www.Davidrumsey.com launched with now over 14,800 historic maps served online. One major collection features unique overlays of the original maps from the Lewis & Clark expedition.	**2002:** Google acquires Picasa, a digital photo management company (http://picasa.google.com/) and publicly offers stock options **2004:** Google acquires Keyhole Corp, a digital and satellite imaging company **2005:** Google Maps and API launch (birth of Mash-ups) Google acquires Keyhole and rebrands as *Google Earth*	**2006:** Microsoft release Virtual Earth 3D (browser-based virtual globe) http://local.live.com

(continued)

Figure 1.1. Timeline of events, publications, dissertations, and innovations related to GIS integration in K–12 schools.

	1980–1990	1990–1995	1996–2000	2001–2005	2006–2010
				Google launches *Google Maps* (http://maps.google.com/) Google launches *GoogleEarth* (http://earth.google.com/) Google Maps are connected to mobile phones	
State of GIS	**1985:** ESRI's *ArcInfo* released ESRI had several city and state contracts to develop GISystems **1988:** National Center for Geographic Information and Analysis (NCGIA) founded. A consortium of universities with a mission of advancing geographic information research, developed a *Core Curriculum* in 1990 (1998 and 2000 versions now at: http://www.ncgia.ucsb.edu/pubs/core.html)	**1995:** ESRI's *ArcView 2.1* released, for both Power MacIntosh and pc platforms **1996:** Berkley Geo-Research Group releases their *Geodesy* project, now at: http://www.bgrg.com/geodesy/page3.htm **1997:** ESRI's *ArcView 3.0* released *Map Objects* and *ArcExplorer* released ESRI's *GIS Store* for maps, software and publications is launched ESRI's *Virtual Campus* is launched **1998:** ESRI launches GeoChallenge in conjunction with National Geographic Society	**2000:** (www.geographynetwork.com), a collaborative and multiparticipant system for publishing, sharing, and using digital geographic information on the Internet. *ArcGIS 8.1* released *ArcExplorer 3.1 Java Edition* released. ESRI hosted its first Education User Conference immediately preceding the Twenty-first Annual International User Conference in San Diego, California. More than 400 K–12 teachers, college and university instructors, school administrators, librarians	**2001:** ESRI launches *MapObjects 2.1* technology—the largest, most comprehensive rebirthing of GIS technology in the history of the industry. **2002:** AAG launches www.aag.org/sustainable MyCOE Launch *My Community, Our Earth* **2003:** ESRI.com launched the Lewis and Clark Web site to commemorate 200 years of geographic exploration and mapping.	**2006:** ESRI releases *ArcWeb Explorer* (GIS in a browser) (http://www.esri.com/software/arcgis/explorer/index.html) **2007:** Kerski and Baker compile an annotated list of online GST tools: http://gis2.esri.com/industries/education/arclessons/search_results.cfm? **2007:** Baker and Kerski launch new http://edcommunity.esri.com , new products that are Mac accessible are accessible. ArcWebExplorer (http://www.arcwebservices.com/awx/index.jsp) and a new edition of *ArcExplorer Java*

MapInfo founded

IDRISI project, early GIS Clark University

GRASS (open source) GIS introduced by US Army Corps of Engineers

and museum professionals shared their experiences using GIS in the classroom for research and public access.

ESRI President Jack Dangermond donates software to develop GIS educational program for secondary school students and instructors in 43 secondary education centers and 16 teacher training centers.in Uruguay

ESRI provided teacher training to Iraqi nationals from the Ministries of Planning, Mines, Housing & Construction, and Water Resources to support redevelopment efforts.

2003: ESRI launches *ArcLessons* (now located at: http://gis2.esri.com/industries/education/arclessons/arclessons.cfm)

Edition for Educators (AEJEE http://edcommunity.esri.com/software/aejee/) with increased functionality for both Windows and Mac users was also introduced. As well, new 'Widgets for MacOS X' developed by Tom Baker(http://edcommunity.esri.com/apps/widgets/

Figure 1.1. Timeline of events, publications, dissertations, and innovations related to GIS integration in K–12 schools (continued).

	1980–1990	1990–1995	1996–2000	2001–2005	2006–2010
Academic & Industry Publications		**1995:** TERC: First National Conference on the Educational Application of Geographic Information Systems (EdGIS) Report. TERC.	**1997:** Alibrandi: Thinking Spatially, Acting Locally, *Green Teacher.* **1997:** Audet & Paris: GIS Implementation model for schools: Assessing the critical concerns, *Journal of Geography.* **2000:** Ludwig & Audet, Eds. *GIS in Schools.* ESRI Press. **2000:** O'Looney, J. *Beyond Maps: GIS Decision making in local government.* ESRI Press.	**2001:** Green (Ed.), *GIS: A Sourcebook for Schools.* **2002:** Malone, Palmer & Voigt, *Mapping Our World.* ESRI Press **2003:** Alibrandi, *GIS in the Classroom.* Heinemann **2003:** Zannelli-English, *Community Mapping.* ESRI Press **2003:** Bednarz & Baker, (Eds.), Research on GIS in Education (special issue), *Journal of Geography.* **2005:** Baker, Internet-Based GIS Mapping in support of K–12 education. *Professional Geographer.* (see: www.blackwell-synergy.com/links/doi/10 **2005:** NRC, *Learning to think spatially: GIS as a support system in the K–12 curriculum.*	**2006:** NRC, *Learning to Think Spatially* **2006:** NRC, *Beyond Mapping* **2007:** Sinton, D. & Lund, J., (Eds.) *Understanding Place: GIS and Mapping Across the Curriculum* **2007:** Google launches its *Sightseer* newsletter for users of Google Earth (see: http://bbs.keyhole.com/ubb/z0302a1700/june07.htm)

Ph.D. Dissertations	For **1990–2002**, see: http://kangis.org/research/abstracts/ which also lists Masters theses on similar themes	**1990** Brown, Malcom H. *The geographic information system in a junior high school environment* **1993** Audet, Richard *Developing a theoretical basis for introducing geographic information systems into high schools: Cognitive implications* Weller, Kay *The appropriateness of GIS instruction in grade six for teaching Kansas water resources* **1994** Paladino, Steven *A role geographic information systems in the secondary schools: An assessment of the current status and future possibilities*	**1999** Sanders, Raymond *GeoTeks: Using Instructional Technology to Implement The Texas Essential Knowledge and Skills for Social Studies at the Middle Level* Donaldson, Daniel P. *An evaluation of multimedia technology in geography education: A case study of two sixth-grade classes in Ohio* **2000** Wigglesworth, John C. *Spatial problem-solving strategies of middle school students: Wayfinding with geographic information systems* Kerski, Joseph J. *The implementation and f geographic information systems technoleffectiveness oogy and methods in secondary education* Olsen, Timothy P. *Situated student learning and spatial informational analysis for environmental problem*	**2001** Crabb, Kevin D. *Case study of geographic information system integration in a high school world geography classroom* Abbott, Thomas *An evaluation of a geographic information system software and its utility in promoting the use of integrated process skills in secondary science* Kaplan, Danielle E. *Mental model development and reasoning about a causal system in a computer-based inquiry environment* **2002** Baker, Thomas R. *The Effects of Geographic Information System (GIS) Technologies on Students' Attitudes, Self-Efficacy, and Achievement in the Middle School Science Classroom*	**2005** White, Shannon *Geographic Information Systems (GIS) and Instructional Technology (IT) Diffusion: K-12 Student and Educator Conceptualizations* http://www.lib.ncsu.edu/theses/available/etd-09092005-050901/

Figure 1.1. Timeline of events, publications, dissertations, and innovations related to GIS integration in K–12 schools (continued).

1980–1990	1990–1995	1996–2000	2001–2005	2006–2010
			Doering, Aaron *Systems to Foster Inquiry Learning*	
			2003	
			Hagevik, Rita *Creating Science Learning Tools Using Geographic Information*	
			Shin, Eui-kyung *Using geographic information system to enhance elementary students' understanding of geography and history: A case study*	

[a] All of the interviewees cited ESRI's decision to hire a K–12 educator as the watershed event of GIS in K–12. Having worked with NGS, Fitzpatrick had worked with an extensive national teacher network.

[b] Interviewee Joseph Kerski stated that these institutes were the most significant in impact as they led to subsequent institutes (1999–2006) led by himself, Fitzpatrick Baker and Anita and Roger Palmer reaching some 200 educators.

Concerning science education, Robert Tinker's (1992) work exposed the possibilities of representing data with digital maps in many curricular areas. He described the power, flexibility and intrigue that maps offered to fourth, fifth, and sixth grade students studying various aspects of the environment as a part of the KidsNet Project. Tinker also described the significance of kids "ground truthing" data, whereby students, using a map, verify the attribute data of a map or satellite image. He suggested that these data confirmations make the processes of ground truthing and mapping "alive and immediate, providing motivation....for mastering experimental techniques, and for pursuing detailed investigations of interactions" (p.42). Furthermore, Tinker noted "GIS software provides a critical link between the immediate and personal level of field observation and global effects and concerns" (p.42).

The successful implementation of any systemic movement requires support, organization, and communication far beyond the classroom walls. Many events, organizations, grant programs, and companies have provided a variety of unique support for the growth and continued success of GIS in K–12 education. The chronology below briefly describes some of these efforts that have had the most substantial impact upon GIS in K–12 education.

1986

Hosted and implemented in 1986 by the National Geographic Society, the Geographic Alliance is a network that aims to bring geography teachers together to promote scholarly communications and activities. Principally, each state has a local alliance, primarily tasked with promoting geographic literacy for all people. Although activities vary from state to state, the alliances strongly support GIS training for teachers, idea sharing through newsletters, and assist with forming collaborative partnerships for data and further teacher training. The National Geographic Society is also the sponsor of Geography Awareness Week. Held in November of each year, the week before Thanksgiving, this program recognizes the importance of geography to students and the public. Since 2000, GIS Day has fallen on the Wednesday of Geography Awareness Week. With its own website, www.gisday.com, GIS Day is a day when professional GIS agents and technicians volunteer to demonstrate GIS to the public, and in particular, outreaching to schools and K–12 students.

1988

In 1988, the National Center for Geographic Information and Analysis (NCGIA) was founded from a consortium of U.S. universities with a mission of advancing geographic information research. This mission, indirectly included education and public outreach activities designed to help meet the increasing demand for trained GIS professionals. Two years after its inception, NCGIA released the *Core Curriculum*, intended to provide a scope and sequence for GIS education at the undergraduate level. Despite the pioneering efforts of the *Core Curriculum*, it was largely regarded as product that was hurried to market and as a result lacked appropriate pedagogical, cognitive, or affective components. The *Core Curriculum* reportedly was analogous to reading a "military training manual" (Audet, 1993, p. 7). While some consideration was given to adapting this curriculum to the high school level, it was largely regarded by classroom teachers as too dense and too technical for supporting classroom instruction (Palladino, 1998). As a result, the Secondary Education Project (SEP) was launched in 1992 with the dual purposes of identifying existing GIS activities for schools and creating new ones. SEP I consisted of a prototype teacher institute, held for the development, testing, and dissemination of K–12 education materials. NCGIA was unsuccessful at immediately securing funds for the continuation and development of the SEP I program. The results of the early program are likened to a series of compromised solutions, constrained time frames, and hints of waning pedagogical clarity (Audet, 1993).

1989

Following the discovery of the RMS Titanic in 1989, the JASON Project used the Internet, printed curriculum, video, and teleconferencing technologies to bring explorations in science, technology, math, and social studies to students. In this project, maps and GIS technology are used to support general student exploration of a yearly theme. In many cases, GIS technologies are also used by project staff and presented to students for enhancing the unit of study.

1991

The Association for Geographic Information's (AGI) 1991 Conference consisted of an organized discussion related to GIS in schools (Palladino, 1994). These discussions later evolved into minor inclusions in the subsequent year's AGI yearbook. Isolated examples of GIS activities in schools were also occurring abroad in Austria, Canada, and in the US in North Carolina, Oregon, and Virginia (Palladino, 1994).

1992

The Center for Image Processing in Education (CIPE) is a nonprofit organization founded in 1992 that attempts to train teachers and students in the use of data visualization tools, including image processing packages and geographic information systems. CIPE has also produced extensive training manuals and curriculum that follow approaches similar to Project

Based Learning. To date, CIPE has over 30 certified instructors who have trained in excess of 3,500 teachers in image processing or geographic information systems. CIPE uses *ArcView* GIS software for teachers and students (CIPE, 2001).

ESRI releases *ArcView* (see: http://www.esri.com/company/about/timeline/flash/index.html for the company time line).

1994

The development of any field within academia or industry is typically marked by a surge of professional meetings and conferences; GIS in K–12 education is no different. In January of 1994, the first annual conference on the pre-college educational applications of Geographic Information Systems was held at the National Geographic Society. Heralded as EdGIS, the conference was a great success and has grown substantially in subsequent years with participants from education, the cognitive sciences, geography, GIS, remote sensing, government, and industry.

Global Learning and Observations to Benefit the Environment (GLOBE) was developed in 1994 for students and teachers to collect and report a variety of environmental data. Visualizations and global data sets are available for schools to download and further analyze with spreadsheets and desktop GIS (GLOBE, 2001). ESRI's *ArcVoyager* software has special adaptations making it easier for classrooms to load GLOBE data.

Initiated in 1994, the Collaborative Visualization Project (CoVis) at Northwestern University gathered students and teachers together in collaborative data collection and visualization, using specialized visualization software created by the project. The Technology Education Research Consortium's (TERC) 1994 National Science Foundation program entitled Mapping Our City had middle-school students using simulations, geographic information systems, and other applications. The intent of the project was to assess the value of such systems in science classrooms, allowing students to construct representations of local spatial phenomena (McWilliams & Rooney, 1997). Project researchers concluded from the two-year effort that GIS helps students discover relationships among variables, simpler GIS technology encourages open-ended explorations of data, and maps help students focus on the spatial nature of data (McWilliams & Rooney, 1997).

1995

Located at the University of Washington, the Nature Mapping environmental education program encourages student data collection of animals and habitats. Started in 1995, students record environmental monitoring information, mail it to project staff, and resulting maps are generated.

Teaching materials and conferences are an integral part of the program (Dvornich, Tudor, & Grue, 1995).

The Visualizing Earth (VisEarth) project, a 1995 collaboration between Pennsylvania State's GeoVista program and the Technology in Education Research Consortium (TERC), had middle school students analyzing remotely sensed and aerial photography data. The project included a cooperative partnership with the NASA-sponsored EarthKAM, a movable camera within a spaceship controlled by students across the Internet. The outcome of this effort has produced some of the latest studies in spatial cognition and recommendations for teaching. Areas of research include scale, point of view, 3-dimensionality, symbolic representation, and change over time (Barstow, Frost, Liben, Ridge, Souveiney, 1999).

1996

The second EdGIS was held in 1996, with a third conference in 1997. It was during EdGIS '96 that a substantial research theme began emerging, where the task of GIS education was subdivided to allow for greater depth of inquiry. These educational subdivisions, largely guided by the NCGIA education division, fit into a framework of pedagogical, curriculum, software, and cognitive issues (NCGIA, 1996).

The Berkley GeoResearch Group (BGRG) created an *ArcView* GIS extension called *Geodesy* for K–12 schools. The application, initially developed in 1996, is now commonly used in 86 middle schools across the United States. Geodesy, like ESRI's *ArcVoyager*, is a streamlined application allowing for more immediate access for student use, minimizing the initial learning curve. Geodesy, emphasizing remote sensing and GIS technologies, is built upon *ArcView 2.1* and allows for a customized local data set and curriculum-based Graphical User Interface. The interface of the package is designed for its singular audience of K–12 students, benchmarked against the Geography for Life Standards, a scope and sequence for K–12 geography education (Radke, 1996).

The WorldWatcher program, released in 1996, includes data, software, lesson plans, and tutorials for K–12 use. The *Greenhouse Effect Visualizer* and the Global Warming curriculum package are products of the continued efforts of the WorldWatcher development (Edelson, Brown, Gordin, Griffin, & Griffin, 1999; Gordin, Edelson, Gomez, 1996). The most recent work from WorldWatcher is in the form of *My World GIS*, a cross-platform Java-based GIS for middle school to college learning applications (see Edelson *et al*, this volume).

Launch of Mapquest.com (http://company.mapquest.com/corporate/2.html)

1997

The Kansas Collaborative Research Network (KanCRN) was created in 1997 as an Internet-based network of schools designed to facilitate student research in the natural sciences. Unlike programs before it, KanCRN involved students as researchers, not simply research assistants. In this way, KanCRN created online environments capable of allowing students to collect, contribute, and analyze data. The use of GIS in the organization, while rooted in *ArcView* GIS, quickly extended into Internet-based mapping and analysis.

1998

In 1998, several conferences related to GIS in education were held. The first International Conference on GIS Education was held in Ypsilanti, Michigan, and stressed "Expanding GIS Education Locally and Globally through Technology and Outreach." Key areas of the conference included the expansion of collaborations between GIS and education communities, the use of problem-centered GIS instruction, and the use of GIS and the Internet. Two other conferences were also held to explore the potential for GIS inclusion in education. GIS Education: A European Perspective (EUGISES98) and the Interoperability for GIScience Education each reinforced indications that a formalization had transpired between GIS and education at both the K–12 and collegiate levels (www.gisday.com).

1999

World Resources Institute (WRI) collects and distributes large amounts of biodiversity and environmental data. In the fall of 1999, WRI and ESRI published an *ArcView* GIS extension for schools called *DataScape*. The application contains over 450 data layers that students and teachers can use independently or to which they can add their own class-collected data (WRI, 2001).

Project VISM (Visualizations in Science and Mathematics), located at James Madison University, is a summer institute program for middle and high school teachers interested in using data visualization technologies in the classroom. This three-year NSF program is set to conclude during 2002 after its final summer of teacher training in image processing, GIS, molecular modeling, and simulations. With continued funding, the program is developing activities for its constituent teachers.

GIS Day, as a collaboration between ESRI, the National Geographic Society, and the Association for Geographic Information, is held annually as a part of Geography Awareness Week. The intent of GIS Day is to educate students and the general public about GIS. Typically, GIS specialists from business and government visit classrooms and demonstrate the technologies, lead tours, or assist in other ways related to GIS awareness. Aside from these traditional GIS Day activities, Internet web casts and map galleries have been created to support student and teacher awareness (www.gisday.com).

ESRI has provided countless hours of training to K–12 teachers during workshops, institutes, and conferences. In 1999, the ESRI K–12 Schools and Libraries division united a core of K–12 educators that were skilled in GIS. This group was to become the flagship team of teacher-trainers in GIS for schools, known as the ESRI K–12 Authorized Teaching Program (ATP). Subsequent teams of ATP teachers were created in 2000 and 2001. It is expected that these ESRI K–12 ATP teams will lead the way in promoting awareness and education in the use of GIS in the classroom. Presently, this group uses a common content scope and sequence (ESRI, 2001c).

ESRI's Community Atlas project was initiated in 1999, to involve classrooms describe their community through student-authored maps and narratives. Throughout the school year, classes would work on their project, culminating in a nation-wide competition between schools and classrooms. Currently, there are over 30 publicly viewable projects submitted to the Community Atlas from the first two years of activity (ESRI, 2001b).

2000

In 2000, the Educational Applications of GIS Conference was hosted at California State University, San Bernardino. The conference lasted three days, with 280 attendees from 8 countries and 40 states. Plenary sessions addressed growth in the GIS industry, education, visual modeling, and online digital libraries.

The Virtual Immersion in Science Inquiry for Teachers (VISIT), a three-year NSF project seated at Eastern Michigan University offered Internet-based support for high school teachers through scientific inquiries into local issues, using visualization technologies for data analysis.

As a part of the Earth Observing System's Education Center, the Lewis and Clark Education Center (LCEC) is attempting to provide students and educators with a historical and present-day vision of the natural and cultural aesthetics of the Lewis and Clark Trail. The program, founded in 2000, uses multimedia and geospatial data to create a rich context of understanding. Through the use of GIS, global positioning systems, and a variety of remotely sensed data sources, the center is creating animations, fly-throughs, and a vast data and document archive. The center sponsored a number of events in 2003, the bicentennial of the expedition, including a national conference and trail trips (Philp, 2001).

2001

During 2001, two conferences were held to support and foster the development of GIS in K–12 education. In San Bernardino, California, a second Educational Applications of GIS was held at California State University, with fewer than 25 attendees. A few weeks later, ESRI sponsored a GIS in Education conference as a pre-conference event to the regular ESRI User Conference. Held in San Diego, California, the conference site hosted over 460 educators interested in sharing ideas, attending hands-on workshops, and exploring the most beneficial ways to integrate GIS in K–12 curricula. The conference was the largest GIS in education event ever held with over 160 presentations by teachers and researchers, from elementary to collegiate levels. Each year since 2003, ESRI has hosted its "EdUC" Educational User Conference in the days preceding the massive ESRI conference.

The NSF-funded program, "Extending Scientific Inquiry with GIS" launched initial development of teacher preparation materials for training K–12 educators in the mechanics of GIS technology, the pedagogical methodologies for inclusion in a standards-based classroom, and spatial reasoning strategies for the science classroom. The initiative, centered at the University of Kansas, is currently planned for three years with final products including field-tested teacher preparation materials and PBL-GIS curricula.

CONCLUSIONS AND IMPLICATIONS

As our key players have indicated, there are paths of the GIS in education landscape still being forged by individuals, projects and initiatives in various school districts, states, and nations. This book is an effort to bring at least those involved in developing GIS in social studies together to begin connecting those paths into a more coordinated network. The analogy is apt because we tend to make connections over the landscape from streams to paths to roads to highways. The landscape thus becomes connected originating from the hydro-logic to a constructed network mirroring the bio-logic of our neural pathways, into a coherent map of the landscape.

As this book was coming press, the 9th ESRI EdUC (Education Users Conference) was being held in San Diego with the newly expanded team. Unveiled for educational users were several new 'roll-out' products—three new online tools for interfacing with IGIS. One of the new tools, ArcGIS Explorer (http://www.esri.com/software/arcgis/explorer/index.html), looks much like Google Earth. From edcommunity.esri.com, new products that are Mac accessible are available. ArcWebExplorer (http://www.arcwebservices.com/awx/index.jsp) and a new edition of ArcExplorerJavaEdition for Educators (AEJEE: http://edcommunity.esri.com/software/aejee/) with increased functionality for both Windows and Mac users were also introduced. Additionally, new widgets for MacOS X and audio support have been developed by Tom Baker (http://edcommunity.esri.com/apps/widgets/). Clearly, the industry sees the need to better match the educational environment and has made new commitments and tools toward that goal.

ESRI's Joseph Kerski also presented at the 5th Annual International Society for Digital Earth (ISDE) conference, co-sponsored by ESRI and Google (http://www.isde5.org/). The conference's history is worth reading to get a sense of the prescience and future of digital earth technologies and where the may be headed (http://www.isde5.org/about_digitalearth.htm).

Throughout the following chapters, that landscape will begin to unfold and expand. With the 'critical mass' of efforts, with the facilitating technologies, within the chronology of the coming years, we as authors and co-editors hope to reflect back on this collection as a significant component in the diffusion of GIS in education, particularly as an STS application, analysis and preferably community problem-solving. As we trace these paths led by the book's trusted leaders on this journey, we hope that the temporal, contextual and social history we have tried to illustrate, much as illustrations in a map's margins, help to frame the trajectory of the diffusion of this innovation of GIS in social studies education.

REFERENCES

Alibrandi, M. (2001) Mapping life and society; Integrating geographic information systems into social studies. *Curriculum-Technology Quarterly, 11*(1), 1–3.

Alibrandi, M. (2003). *GIS in the classroom: Using geographic information systems in social studies and environmental science.* Portsmouth, NH: Heinemann.

Alibrandi, M., Beal, C., Wilson, A., & Thompson, A. (2000). Reconstructing a school's history using oral histories and GIS mapping. *Social Education, 64*(3), 134–140.

Alibrandi, M., Bull, P., Sarnoff, H., & White, S. (2004, November). New Landscapes of Learning for Democracy with GIS. Presented at the Annual Conference of the National Council for Social Studies, Baltimore, MD.

Alibrandi, M., White, S., Duke, B., & Hagevik, R. (2001, June). *GIS in teacher education: Notes from the field II.* Presented at the Annual ESRI EdUC Educational user conference, San Diego, CA: Environmental Systems Research Institute.

Audet, R.H. (1993). *Developing a theoretical basis for introducing geographic information systems into high schools: Cognitive implications.* Unpublished doctoral dissertation, Boston University.

Audet, R., & Ludwig, G., Eds. (2000). *GIS in Schools.* Redlands, CA: ESRI Press.

Baker, T. (2005). Internet-based GIS mapping in support of K–12 education. *The Professioanl Geographer, 57*(1), 44–50.

Baker, T. R. (2002). *The effects of geographic information system (GIS) technologies on students'attitudes,self-efficacy, and achievement in middle school science classrooms.* Unpublished doctoral dissertation, University of Kansas. Retrieved from http://tbaker.com/tbaker/academics/papers/dissertation/public_release/trbaker_dissertation.pdf

Barr, R. D., Barth, J. L., & Shermis, S. (1977). *Defining the social studies.* Washington, DC: NCSS.

Barstow, D., Frost, E., Liben, L., Ridge, S., & Souveiney, R. (1999). *Final report to the National Science Foundation: Visualizing Earth.* Retrieved August, 2001, from http://visualizingearth.ucsd.edu.

Bednarz, S. W. (1999). *Impact and success: Evaluation of the GIS Institute for teachers.* Proceedings, ESRI User Conference, San Diego, CA. Available at http://gis.esri.com/library/userconf/proc99/proceed/papaers/pap895/p895.htm

Center for Image Processing in Education (CIPE). (2001). *The center for image processing in education.* Retrieved August, 2001 from http://www.airedesign.com/CIPESite/

Davis, O. L. (1981). *The social studies: 80th yearbook of the national society for the study of education, Part II.* Chicago, IL: University of Chicago.

Dewey, J. (1916). *Democracy and education: An introduction to the philosophy of education.* New York: Free Press.

Dvornich, K. M., Tudor M., & Grue, C. E. (1995). Naturemapping: Improving management of natural resources through public education and participation. *Wildlife Society Bulletin, 23* (4), 609–614.

Edelson, D., Brown, M., Gordin, D., & Griffin, D. (1999). Making visualization accessible to students. *GSA Today, 9* (2), 8–10.

Environmental Systems Research Institute (ESRI). (2001a). *Higher Education.* Retrieved August, 2001, from http://www.esri.com/industries/university/colleges.html

Environmental Systems Research Institute (ESRI). (2001b). *The Community Atlas.* Retrieved August, 2001, from http://www.esri.com/industries/k-12/atlas/

Environmental Systems Research Institute (ESRI). (2001c). *ESRI authorized teaching program for K12 educators.* Retrieved August, 2001, from http://www.esri.com/industries/k-12/atp/

Freire, P. (1970). P*edagogy of the oppressed.* New York: Continuum.

Friedman, T. (2005). *The world is flat: A brief history of the twenty-first century.* New York: Farrar, Strauss & Giroux.

GLOBE (2001). *Global learning and observations to benefit the environment.* Retrieved October 15, 2007, from http://www.globe.gov/fsl/html/templ.cgi?about&lang=en&nav=1

Golledge, R. G., & Stimson, R. J. (1997). *Spatial behavior: A geographic perspective.* New York: Guilford.

Gordin, D. N., Edelson, D. C., & Gomez, L. M. (1996, July). Scientific visualization as an interpretive and expressive medium. In D. C. Edelson & E. A. Domeshek

(Eds.), *Proceedings of the Second International Conference on the Learning Sciences*, pp. 409–414.

Harding, S. (1986). *The science question in feminism.* Ithaca, NY: Cornell.

Hartoonian, H. M. (1996). The price of civilization: Competence and constant vigil. Presidential Address delivered at the 75th Annual Conference of the National Council for the Social Studies, Chicago, IL, *Social Education,* 6–8.

Kerski, J. J. (2000). *The implementation and effectiveness of geographic information systems technology and methods in secondary education.* Unpublished doctoral dissertation, University of Colorado at Boulder.

Kosslyn, S. (1994). *Image and brain.* Cambridge, MA: MIT Press.

Kwan, M-P. (2002). Feminist visualization: Re-envisioning GIS as a method in feminist geographic research. *Annals of the Association of American Geographers, 92* (4), 645–661.

Lynch, K. (1960). *Image of the city.* Boston: South End Press.

Malone, L., Palmer, A., & Voigt, C. (2005). *Mapping our world: GIS lessons for educators.* Redlands, CA: ESRI Press.

McWilliams, H., & Rooney, P. (1997, March). *Mapping our city.* Conference presentation at the Annual Meeting of the American Educational Research Association. Chicago, IL.

Milson, A. J., & Alibrandi, M. (In press). Critical map literacy and internet-based geographic information systems: The spatial dimension of civic decision-making. In M. Berson & P. VanFossen (Eds.). *The Electronic Republic? The Impact of Technology on Education for Citizenship.* Purdue University Press.

National Center for Geographic Information Analysis (NCGIA). (1996). EDGIS '96 research agenda meeting. Retrieved March, 1999 from http://www.ncgia.ucsb.edu/education/projects/SEP/EDGIS96/EDGIS96.html

NCSS. (n.d.) *NCSS curriculum standards for social studies: Executive summary.* Retrieved June 12, 2007, from http://www.socialstudies.org/standards/execsummary/

National Research Council. (2005). *Learning to think spatially: GIS as a support system in the K–12 curriculum.* National Academies Press.

O'Looney, J. (2000). *Beyond Maps: GIS decision-making in local government.* Redlands, CA: ESRI Press.

Palladino, S. (1994). *A role geographic information systems in the secondary schools: An assessment of the current status and future possibilities.* Retrieved March, 1999 from http://www.ncgia.ucsb.edu/~spalladi/thesis/title.html

Palladino, S. (1998). *K–14 GIS: A review of eight years of NCGIA efforts.* Retrieved March, 1999 from http://www.ncgia.ucsb.edu/~spalladi/documents/ESRI.html

Philp, A. (2001). *The Lewis and Clark education center.* Retrieved August, 2001 from http://lewisandclarkeducationcenter.com

Poiker, T. K. (1985). Geographic information systems in the geographic curriculum. *The Operational Geographer, 8,* 38–41.

Radke, S. L. (1996). *GEODESY: An educational series for youth.* Retrieved February 1999 from http://www.esri.com/library/userconf/proc96/T

Rogers, E. (2003). *Diffusion of Innovations (5th ed.).* New York: Free Press.

Sinton, D., & Lund, J. (2007). *Understanding place: GIS and mapping across the curriculum.* Redlands, CA: ESRI.

White, S. (2005). *Geographic information systems (GIS) and instructional technology (IT) diffusion: K–12 student and educator conceptualizations*. Unpublished doctoral dissertation, North Carolina State University. Retrieved April 23, 2007, from http://www.lib.ncsu.edu/ theses/available/etd-09092005-050901/

World Resources Institute (WRI). (2001). *World resources institute education center.* Retrieved August, 2001 from http://www.wri.org/wri/enved/

CHAPTER 2

THE STATUS OF GEOSPATIAL TECHNOLOGIES IN U.S. HIGH SCHOOL GEOGRAPHY STANDARDS

Andrew J. Milson
University of North Texas

Jennifer A. Roberts
University of North Texas

The chorus of voices calling for more focused attention on geospatial technologies in K–12 schooling grew more voluble with the publication of two reports in 2006. In *Learning to Think Spatially*, the National Research Council (NRC) recommended that K–12 educators foster spatial literacy across the curriculum (National Research Council, 2006). The report authors identified spatial thinking as a significant, yet overlooked and under-taught, skill that is essential to citizenship, workforce preparation, and conceptual understanding in a variety of disciplines. Furthermore, they described how both high and low tech support systems should be employed when teaching spatial thinking to students in K–12 contexts. In addition to the efforts of the NRC, the results of the work of a committee formed by the University Consortium for Geographic Information Science (UCGIS) were published in 2006 as the *Geographic Information Science & Technology (GIS&T) Body of Knowledge* (DiBiase et al.,

Digital Geography: Geospatial Technologies in the Social Studies Classroom, pages 39–59
Copyright © 2008 by Information Age Publishing
All rights of reproduction in any form reserved.

2006). The *GIS&T Body of Knowledge* consists of ten "knowledge areas" containing seventy-three units and more than 300 topics. The purpose of the document is to guide curriculum development efforts in higher and continuing education in response to the demand for better preparation of professionals for the expanding geospatial workforce. Although not directly concerned with the K–12 context, the authors of the *GIS&T Body of Knowledge* recognized the significant implications of slow adoption of geospatial technologies in K–12 environments upon post-secondary education. The authors noted, "Too many students who were not exposed to geospatial technologies in primary and secondary schools discover it too late in their college careers to study it in much depth as undergraduates" (DiBiase et al., 2006, p. 19).

There has been much speculation regarding the reasons for the slow adoption of GIS and related technologies in K–12 education. The barriers to GIS implementation most frequently noted in the literature include factors such as the complexity of the software, the limitations of school technology infrastructures, and teachers' lack of familiarity and comfort with GIS for instruction (Baker, 2005; Bednarz & Audet, 1999; Bednarz & van der Schee, 2006; Donaldson, 2001; Gatrell, 2004; Kerski, 2003; Milson, DeChano, Bunch, Caito, & Qiu, 2005). Sparse attention has been given to the presence or absence of GIS in state standards as another barrier to widespread implementation. Given that K–12 education is driven significantly by curriculum standards and standardized testing, we searched the standards for each state in the United States to determine which states give explicit attention to geospatial technologies in their standards for high school geography. Although both the NRC and the UCGIS reports make it clear that spatial thinking and GIS&T knowledge spans multiple disciplines, we focused our search on high school geography standards because this is the most likely disciplinary home for GIS within high school curriculum standards. We were most interested in those standards that explicitly mention GIS, GPS, or analysis of geographic data and information with technology. Although these topics might be implicit within the standards of other fields, the geography curriculum is the most likely home for such goals. Since geography is not required for graduation in every state, we also sought to determine which states both require geography and include reference to geospatial technologies in geography standards. Thus, our search was guided by five questions:

1. In which states do the high school geography standards refer to Geographic Information Systems?
2. Which states' high school geography standards mention the Global Positioning System?

3. In which state standards is there reference to the analysis of geographic data and information using technological support systems?
4. Which states require geography for high school graduation?
5. What is the language used in the standards when reference is made to GIS, GPS, or the analysis of geographic information?

GEOGRAPHIC INFORMATION SYSTEMS IN STATE GEOGRAPHY STANDARDS

We identified eighteen states that refer to GIS explicitly in their state standards for geography (see Table 2.1), yet the suggested application of GIS in these state standards varies widely. In some states students are only expected to be aware of GIS, while in a few others GIS is described as a tool for analysis of geographic data and students are expected to use GIS to investigate geographic questions. The latter is evident in the standards for students in Minnesota, Indiana, and Texas. Minnesota students are expected to "use maps, globes, geographic information systems, and other databases to answer geographic questions at a variety of scales from local to global" (Minnesota Department of Education, 2005). Minnesota students and teachers are also provided specific benchmarks and examples related to this standard.

1. Students will demonstrate the ability to obtain geographic information from a variety of print and electronic sources. Examples: Atlas, World Wide Web, topographic maps, Geographic Information Systems, databases, aerial photos.
2. Students will make inferences and draw conclusions about the character of places based on a comparison of maps, aerial photos, and other images. Example: Make a land use map of a local area.
3. Students will demonstrate the ability to use geographic information from a variety of sources to determine feasible locations for economic activities and examine voting behavior. Examples: Fast food restaurant in local community, a good place to found city, put a church or military installation, locate a solid waste disposal site, locate a feedlot, voting in presidential elections (Minnesota Department of Education, 2005).

The standards for students in Indiana also focus on the application of GIS for geographic questions. Indiana geography students are expected to "use maps, globes, atlases, and grid-referenced technologies, such as remote sensing, Geographic Information Systems (GIS), and Global Positioning Systems (GPS) to acquire and process information about people,

TABLE 2.1
Geospatial Technology Standards in the Geography Curriculum of U.S. States

State	GIS Standard	GPS Standard	Analysis of Geographic Information Standard	Geography Required for High School Graduation
Alabama				
Alaska			•	
Arizona	•		•	
Arkansas[a]				
California				
Colorado	•		•	~
Connecticut				~
Delaware	•	•		
Florida	•			
Georgia				
Hawaii				
Idaho	•	•		
Illinois			•	
Indiana	•	•	•	
Iowa[b]				
Kansas	•	•	•	
Kentucky				~
Louisiana[c]				
Maine				
Maryland				
Massachusetts				~
Michigan				
Minnesota	•		•	•
Mississippi[d]	•	•		
Missouri				
Montana	•		•	
Nebraska				
Nevada			•	
New Hampshire				
New Jersey[e]				
New Mexico	•		•	
New York	•		•	
North Carolina[f]				
North Dakota				
Ohio[g]				
Oklahoma			•	
Oregon				
Pennsylvania	•			

State	GIS Standard	GPS Standard	Analysis of Geographic Information Standard	Geography Required for High School Graduation
Rhode Island				
South Carolina	•			
South Dakota				•
Tennessee	•	•		
Texas	•		•	•
Utah	•			•
Vermont				
Virginia				
Washington[h]	•			
West Virginia				
Wisconsin				
Wyoming				

Note: All data were collected from state department of education websites. Revisions to state curriculum standards posted to websites after August 2006 are absent from this survey. Bullet point (•) indicates presence of standard in this state. Tilde (~) indicates that graduation requirements are determined by local authorities.

[a] Arkansas standards mention GIS in grades 5–8, but not in high school (Arkansas Department of Education, 2000).
[b] Iowa does not have written standards for social studies.
[c] Louisiana standards mention geography tools and technology beginning in grades K–4 and continuing through high school, but there is no specific reference to GIS, GPS, or geographic data analysis (Louisiana Department of Education, 1997).
[d] Beginning with the class of 2012, Mississippi students earning a College Prep Diploma will take one half credit in geography.
[e] New Jersey standards mention computer databases and satellite maps in all grade levels. GIS is referred to in grades 5–8, but not in high school standards (New Jersey Department of Education, 1999).
[f] In North Carolina, there is an elective course entitled Geography in Action that contains significant reference to GIS (North Carolina Department of Public Instruction, 2003).
[g] The Ohio standards refer to GIS in an isolated section on technology. GIS is not integrated into the grade level standards (Ohio Department of Education, 2002).
[h] The Washington standards refer to GIS as a tool for interpretation of data in the grade 8 benchmarks.

places, and environments" (Indiana Department of Education, 2001). Specifically, students are required to:

- Use locational technology (remote sensing, Global Positioning Systems [GPS] and Geographic Information Systems [GIS]) to establish spatial relationships.

- Explain that maps contain spatial elements of point, line, area, and volume, each of which must be digitized differently for incorporation within a GIS.
- Ask geographic questions and obtain answers from a variety of sources, such as books, atlases, and other written materials; statistical source material; fieldwork and interviews; remote sensing; word processing; and GIS. Reach conclusions and give oral, written, graphic, and cartographic expression to conclusions (Indiana Department of Education, 2001).

Furthermore, Indiana educators expect that students will "identify and analyze the human activities that shape Earth's surface, including population numbers, distribution and growth rates, rural and urban land use, ways of making a living, cultural patterns, and economic and political systems. Using grid-based technology, such as remote sensing and GIS wherever possible, they will map the distribution of various human phenomena and look for spatial patterns that the maps reveal" (Indiana Department of Education, 2001).

In addition to Minnesota and Indiana, the Texas standards for high school geography include detailed expectations related to GIS as a tool for higher level thinking. The Texas standards include GIS within the "Social Studies Skills" expectations for geography students as a means for applying "critical-thinking skills to organize and use information acquired from a variety of sources including electronic technology" (Texas Education Agency, 1998a). Specifically, the Texas student is expected to:

- Use historical, geographic, and statistical information from a variety of sources such as databases, field interviews, media services, and questionnaires to answer geographic questions and infer geographic relationships;
- Analyze and evaluate the validity and utility of multiple sources of geographic information such as primary and secondary sources, aerial photographs, and maps;
- Construct and interpret maps to answer geographic questions, infer geographic relationships, and analyze geographic change;
- Apply basic statistical concepts and analytical methods such as computer-based spreadsheets and statistical software to analyze geographic data; and
- Use a series of maps, including a computer-based geographic information system, to obtain and analyze data needed to solve geographic and locational problems (Texas Education Agency, 1998a).

Most of the state standards that mention GIS do not provide the detail given in the Minnesota, Indiana, and Texas standards. Several states give brief mention to GIS in the context of analyzing geographic patterns and solving geographic problems. For example, Arizona students should "analyze how geographic knowledge, skills, and perspectives (e.g., use of Geographic Information Systems in urban planning, reapportionment of political units, locating businesses) are used to solve contemporary problems" (Arizona Department of Education, n.d.). A student in Florida is expected to use "a variety of maps, geographic technologies including geographic information systems (GIS) and satellite-produced imagery, and other advanced graphic representations to depict geographic problems" (Florida Department of Education, n.d.). Similarly, Kansas students should be able to "produce maps and other geographic representations, using data from a variety of sources to answer questions and solve problems (e.g., census data, interviews, geographic information system (GIS) and other databases, questionnaires)" (Kansas State Department of Education, 2005). Finally, the New Mexico standards define a geographic tool as, "a device used to compile, organize, manipulate, store, report, or display geographic information, including maps, gazetteers, globes, graphs, diagrams, aerial photographs, satellite images, geographic information systems, and other computer databases and software" and state that high school students should be able to "analyze and evaluate the characteristics and purposes of geographic tools, knowledge, skills, and perspectives, and apply them to explain the past, present, and future in terms of patterns, events, and issues" (New Mexico State Board of Education, 2001). The standards in Arizona, Florida, Kansas, and New Mexico acknowledge the role of GIS as a tool for problem-solving and suggest that students are to use GIS for this purpose, yet GIS is frequently grouped with several other "geographic tools" and the language used to describe its use is vague when compared with the detail of the Minnesota, Indiana, and Texas standards.

Another group of states mention GIS in their standards, but do so without referring to problem-solving or specific applications of GIS. The Mississippi standards, for example, state that students should demonstrate the ability to "synthesize information from various representations of the earth (e.g., special purpose maps, outer space photos, GPS, GIS, etc.)" (Mississippi Department of Education, 2004). Similarly, Montana requires students to "interpret, use, and synthesize information from various representations of the Earth (e.g., maps, globes, satellite images, geographic information systems, three-dimensional models)" (Montana Office of Public Instruction, 2000). In Utah, students are expected to "collect and interpret geographic data using maps, charts, population pyramids, cartograms, remote sensing, and Geographic Information Systems (GIS)" (Utah State Office of Education, 2002). South Carolina students must "interpret information obtained from maps, aerial photographs, satellite-produced images, and geographic information systems" (South Carolina Department of Education, 2005). Finally, Washington students should be able to "use globes, a variety of map projections, satellite imagery, and Geographic Information System (GIS) data to interpret information from a spatial perspective" (Washington Office of Superintendent of Public Instruction, 2003). The standards for each of these states seem to conceive of GIS as a form of electronic atlas to be used primarily for map interpretation exercises. The notion that GIS is a collection of computer maps is implicit in the standards of these states. Moreover, the standards in these five states tend to group GIS with globes and satellite imagery without distinguishing how GIS can be applied to data management and spatial problem-solving in ways that are unique.

Of the seventeen states that explicitly mention GIS in their state geography standards, four states require what we would classify as *GIS awareness* rather than *GIS application*. For example, high school students in Delaware are only expected to "explain how GPS and GIS are used to gather, portray and analyze data in thematic map form" (Delaware Department of Education, 2001). Idaho students should be able to "apply latitude and longitude to locate places on Earth and describe the uses of locational technology, such as GPS and GIS" (Idaho State Department of Education, n.d.). Likewise, Pennsylvania ninth graders need only explain the uses of geographic tools such as, "geographic information systems [GIS], population pyramids, cartograms, satellite-produced images, climate graphs, access to computer-based geographic data (e.g., Internet, CD-ROMs)" (Pennsylvania Department of Education, 2002). Finally, the expectation for students in Tennessee is that they are able to "explain the characteristics and uses of current geographic techniques in mapping such as geographic information systems (GIS), global positioning systems (GPS), and satellite-produced imagery" (Tennessee Department of Education, 2001). Although these states are recognized in Table 2.1 as those with a GIS standard, the expectations for GIS usage are much more modest than those in states such as Minnesota, Indiana, and Texas. Certainly, awareness of GIS is better than ignorance of GIS, but the power of GIS for engaging students in higher level thinking is lost if they are required to do little more than recognize the acronym.

The Global Positioning System in State Geography Standards

Only six states explicitly mention the GPS in their state standards for geography. Of these six states, only Indiana includes a GPS standard that suggests more than *GPS awareness*. The Indiana standards define the GPS as "a system of satellites and ground stations used to locate precise points on the surface of Earth" and require that students will use the GPS and other technologies to "acquire and process information about people, places, and environments" and "establish spatial relationships" (Indiana Department of Education, 2001). By contrast, the states of Delaware, Idaho, and Tennessee only require that students are able to explain and describe the uses of the GPS. The Kansas benchmarks do not mention the GPS, but it is included within the "High School Instructional Suggestions." In this section, it is suggested that Kansas high school students, "choose and give reasons to use specific technologies to analyze selected geographic problems (e.g., aerial photographs, geographic information systems (GIS), geographic (*sic*) positioning system (GPS), to determine the extent of water pollution in harbors, or the range of deforestation in several locations)" (Kansas State Department of Education, 2005). While the mention of the GPS as a tool for geographic problem solving is noteworthy and the suggestion that students should select and justify the use of appropriate technology applications is laudable, it is unfortunate that the Kansas standards incorrectly label the system as *geographic* positioning instead of *global* positioning.

Although the state of Mississippi also mentions the GPS in its social studies framework, the GPS standard is problematic. Within the section that requires students to, "demonstrate the ability to apply and interpret social studies tools," is the statement that students should "synthesize information from various representations of the earth (e.g., special purpose maps, outer space photos, GPS, GIS, etc.)" (Mississippi Department of Education, 2004). It is unclear how the GPS should represent one of the "various representations of the earth." The GPS is not a representation of the earth in the sense of a map or a satellite image. Strictly speaking, the GPS also is not a source of geographic information but merely a source of geographic data (See Longley, Goodchild, Maguire, & Rhind, 2001). The suggestion that students should synthesize information from maps, photos, and the GPS could be interpreted to mean that students should acquire and analyze various geospatial data using a GIS. This is a very generous interpretation of the language of this standard. It is more likely that the language of this standard, if not ignored completely, would be interpreted to mean something similar to *GPS awareness.* Given that there is no definition of the GPS and the acronym is not spelled out in the standard, we must question whether Mississippi students truly encounter the GPS in a meaningful way in the course of their geographic education.

STANDARDS FOR THE ANALYSIS OF GEOGRAPHIC DATA AND INFORMATION

We recognize that some of the authors of state standards may have shunned the specificity of expectations that explicitly call for the use of a GIS and the GPS. Both of these technologies require investment in hardware and software that may not be realistic within local budget constraints. Yet, we have also noted above that some of the states that call for GIS and GPS in their standards do so in a way that is limited to awareness of these technologies rather than application of these technologies. For these reasons, we surveyed the state standards to determine which states called for some form of analysis of geographic information with technological tools. We sought to identify those states that did not include GIS or GPS in their standards, but who did call for geographic data analysis with technologies. We also sought to identify those states that revealed a comprehensive approach to geospatial technologies in their geography standards.

We identified four states in which the geography standards did not mention GIS or the GPS explicitly, but where the standards did call for analysis of geographic information with technology. In Alaska, for example, students are expected to "use graphic tools and technologies to depict and interpret the world's human and physical systems" and to "use spatial (geographic) tools and technologies to analyze and develop explanations and solutions to geographic problems" (Alaska Department of Education and Early Development, n.d). Similarly, the Illinois Learning Standards call for high school students to "use maps and other geographic instruments and technologies to analyze spatial patterns and distributions on earth" and to "demonstrate how maps, other geographic instruments and technologies are used to solve spatial problems (e.g., land use, ecological concerns)" (Illinois State Board of Education, n.d.). In Nevada, students should be able to "use appropriate geographic tools and technologies to analyze and interpret Earth's physical and human features" and "use a variety of tools and technologies to select and design appropriate forms of maps, graphs, diagrams, tables, or charts to organize geographic information" (Nevada Department of Education, 2000). Finally, Oklahoma students are expected to "use maps and other geographic representations, tools, and technologies to acquire, process, and report information from a spatial perspective" and "apply geographic representations and technologies to depict, analyze, explain, and solve geographic problems" (Oklahoma State Department of Education, 2002). The language of the standards in Alaska, Illinois, Nevada, and Oklahoma indicates an expectation of analysis of geographic information with technologies without explicitly naming GIS or the GPS as a component of the standard. We believe that these standards would be strengthened with the addition of explicit reference to the GPS and analysis with a GIS. In addition to Alaska, Illinois, Nevada, and Oklahoma, nine other states call for the analysis of geographic information with technology tools in their geography standards. These nine states (Arizona, Colorado, Indiana, Kansas, Minnesota, Montana, New Mexico, New York, and Texas) also have an explicit GIS standard, and two of these eight (Indiana and Kansas) include a GPS standard. In sum, we identified only thirteen states where high school geography students are expected to analyze geographic information using technological tools.

Comprehensive Standards for Geospatial Technologies

Five states have standards that we consider to be comprehensive expectations for the use of geospatial technologies in high school geography: Colorado, Indiana, Minnesota, New York and Texas. In our view, comprehensive standards go beyond *GIS awareness* to set expectations for the *application* of geospatial technologies for higher level thinking about spatial patterns and geographic problems. These standards most closely address the recent calls for attention to spatial thinking across the curriculum (NRC, 2006) and the preparation of the workforce and citizenry for the ubiquitous presence of geospatial technology-supported decision making (DiBiase et al., 2006). The standards for Indiana, Minnesota, and Texas were highlighted earlier as exemplars of GIS standards. The standards in these states should be considered comprehensive because they set expectations for students to engage in higher level thinking about geographic problems using technology. Additionally, the standards for students in Colorado and New York are excellent examples of comprehensive geospatial technology standards.

The well-written standards for Colorado students provide a convincing rationale for geography and the use of geospatial technologies for geographic problem-solving. The Colorado standards state that students should, "know how to use and construct maps, globes, and other geographic tools to locate and derive information about people, places, and environments" (Colorado Department of Education, 1995). The following rationale is provided for this expectation.

> Seeing the world geographically requires an understanding of various tools to be able to interpret and make maps; recognize relationships in and between places; make generalizations; and understand the concepts of distance, direction, location, connection, and association. These abilities and concepts are basic to what makes geography unique—the spatial perspective. Maps, globes, photographs, satellite images, and geographic information systems (GIS) are examples of geographic tools. They are essential to portraying, analyzing, evaluating, and predicting human and physical patterns and

> processes on Earth's surface. They play a critical role in helping people make sense of a complex world, and they improve human capacity to move about and plan activities (Colorado Department of Education, 1995).

This rationale statement reflects a deep understanding by the Colorado standards authors of the value and potential for geospatial technologies within geographic education. Beyond knowing about GIS, Colorado students are expected to be able to select "appropriate maps, map projections, and other graphic representations to analyze geographic problems" and to interpret "maps and other geographic tools, through the analysis of case studies and using data" (Colorado Department of Education, 1995).

The New York standards authors cited the 1994 National Geography Standards frequently as they described the expectations for geography education in New York. For example, New York students are expected to "analyze geographic information by developing and testing inferences and hypotheses, and formulating conclusions from maps, photographs, computer models, and other geographic representations" and to "develop and test generalizations and conclusions and pose analytical questions based on the results of geographic inquiry" (New York State Education Department, 1996). The New York standards also provide teachers with helpful examples of geographic analysis and inquiry. One example of geographic problem solving suggests that students:

> Examine a collection of maps, photographs, satellite-produced images, databases, graphs, firsthand accounts, and other sources to generate questions and pose problems for investigation; organize primary and secondary geographic sources to investigate local, national, and international environmental problems and issues; determine the many perspectives that individuals and groups advocate as they seek to resolve these problems or issues; apply a geographic perspective along with others in reaching conclusions on the issues (New York State Education Department, 1996).

A second example provided in the New York standards describes how GIS might be incorporated into the analysis of geographic information. The standards suggest that students:

> Analyze geographic information contained in a spatial database or geographic information system (GIS) to answer questions concerning land use, economic development, population shifts, and transportation networks; use maps showing family income, transportation systems, natural resources, recreation areas, educational and health facilities and other data to develop hypotheses about why some regions prosper and others do not (New York State Education Department, 1996).

These examples suggest a depth to the process of analyzing geographic patterns and problems with technology. This depth is an important element of what we consider to be the comprehensive nature of the geospatial technology standards that are in place in Colorado, Indiana, Minnesota, New York and Texas.

HIGH SCHOOL GRADUATION REQUIREMENTS

If students are not required to take a geography course in high school, then the standards discussed thus far apply to only a fraction of the students in a given state. In the final portion of our survey, we identified only four states where a separate geography course is required for high school graduation (Minnesota, South Dakota, Texas, and Utah). In many states, geography is fused with world history and/or United States history. For example, in California students are required to take, "three years of history/social science, including one year of U.S. history & geography; one year of world history, culture, and geography; and one semester each of American government and economics" (California State Board of Education, 1998). Our review of the standards for fused courses such as these revealed that the geography content is typically overshadowed by history content expectations. In several other states, students must choose between a geography course and a world history course. In Georgia, for example, the requirements state:

> At least three units of credit shall be required in social studies. One unit of credit shall be required in United States history. One unit of credit shall be required from the world studies area, e.g., world history or world geography. World history shall be required for the College Preparatory and College Preparatory with Distinction seals. World geography may be taken to meet the world studies requirement for a Technology/Career or Technology/Career with Distinction seal (Georgia Department of Education, 2006).

It is unfortunate that some states persist in labeling world geography as a course only for those students not intending to go to college. Additionally, we identified four states where local districts determine graduation requirements within state guidelines (Colorado, Connecticut, Kentucky, and Massachusetts). For example, the Kentucky requirements state that three credits of social studies are required and that these credits must, "incorporate U.S. History, Economics, Government, World Geography and World Civilization. Districts and schools can arrange the essential content within the three-credit requirement to best meet their need" (Kentucky Department of Education, n.d.). It is conceivable that some of the localities in these states may require a geography course in high school, but we did not examine local requirements to determine to what degree this is true.

Of the four states where a geography course is required for high school graduation, only Texas requires a full credit of world geography (Texas Education Agency, 1998b). In Utah and South Dakota, students are required to take one half credit of geography (South Dakota Department of Education, 2005; Utah State Office of Education, 2002). The Minnesota requirements are not specific about the exact credit distribution of the social studies requirement, but state that students must take, "3.5 credits, encompassing U.S. history, geography, world history, economics, and government/citizenship - or - 3 credits encompassing U.S. history, geography, world history, government/citizenship and a .5 credit in economics (taught in social studies, business, or agriculture education department)" (Minnesota Department of Education, 2006).

When we compare the five states that have comprehensive standards for geospatial technologies in high school geography with the four states that require this geography course for all high school students, we find that only Texas and Minnesota students are likely to encounter these comprehensive standards on a large scale. For students in Indiana, New York, and Colorado, the excellent standards of their states are merely one option within their high school programs. Although students in Utah and South Dakota are required to take a separate geography course in high school, the attention to geospatial technologies in the standards of these states could be improved. Students in Utah are required to take one half credit of geography and their standards mention GIS, but the Utah standards would be strengthened with the addition of expectations for geographic data analysis using geospatial technologies. South Dakota students are also required to take one half credit of geography, but the standards for this course are vague when addressing geospatial technologies. The South Dakota standards state that high school students performing on the advanced level should be able to, "evaluate and select resources, data services, and geographic tools to generate, manipulate, and interpret information" (South Dakota Department of Education, 2006). There is no explicit expectation in this statement about what forms of data services or geographic tools are to be used. The standards for Utah and South Dakota geography students would be strengthened if they were to adopt the more comprehensive framework evident in the standards in Texas and Minnesota.

STRENGTHENING STATE GEOGRAPHY STANDARDS

Most states have significant room to improve their high school geography standards to better address geospatial technologies. We offer three recommendations to the authors of future state standards that we believe will strengthen their students' knowledge and abilities related to geospatial technologies. First, we recommend that states adopt a comprehensive approach to geospatial technologies. Second, we suggest that states examine the vertical alignment of their programs in geography and social studies to ensure that geospatial technologies are introduced in a sensible and continuous sequence. Finally, we propose revisions to the national standards for geography and social studies to incorporate greater attention to geospatial technologies.

A Comprehensive Approach

In this chapter, we have identified exemplars of comprehensive state standards for geospatial technologies in high school geography. The states of Colorado, Indiana, Minnesota, New York and Texas have developed standards that include explicit attention to GIS and the analysis of geographic data and information with technology. These states emphasize the application—rather than simple awareness—of GIS and geospatial technologies for investigating geographic problems and spatial patterns. Geospatial technologies should not be buried within a list of "geographic tools" as many states have done. There should be separate and explicit attention to geospatial technologies that calls attention to their real-world applications and their distinctiveness when compared with other geographic tools. For example, students should understand that a GIS is more than an electronic atlas. They should understand how geospatial technologies are being applied every day to geographic and social studies issues. Given that a high school geography course is rarely a requirement for high school graduation in the United States, we also recommend that increased attention be given to geospatial technologies in other required social studies courses. Although geography is the most logical home for the study of geospatial technologies, there are many avenues for enhancing the study of history, government, and economics with geospatial technology support. For example, states might include expectations that history students examine the geographic patterns evident in historical census data using an Internet-based GIS (see Radinsky, this volume).

Vertical Alignment of Geospatial Technology Standards

Another important consideration for the authors of state standards is the degree to which the expectations at various grade levels are aligned. Many curriculum developers refer to this as vertical alignment. We found it curious that states such as Arkansas, New Jersey, and Washington mention GIS in the middle school grades, but do not refer to it again in the high school geography course (Arkansas Department of Education, 2000; New Jersey Department of Education, 1999; Washington Office of Superintendent of Public Instruction, 2003). On one hand, it is encouraging that students in these states are potentially exposed to GIS in middle school. On the other hand, the states' standards would be stronger if students were expected to build upon their middle school experiences in high school. With greater attention to vertical alignment, states would produce standards that provide gradual development of knowledge and skills in geospatial technologies so that students would experience these applications continuously over the course of their K–12 schooling.

The Montana standards provide one example of vertical alignment related to geospatial technologies. The Montana standards include expectations for grades 4, 8, and 12. Students in grade 4 in Montana are to, "use appropriate geographic resources (e.g., atlases, databases, charts, grid systems, technology, graphs, maps) to gather information about local communities, reservations, Montana, the United States, and the world" (Montana Office of Public Instruction, 2000). By grade 8, Montana students should be able to, "use appropriate geographic resources to interpret and generate information explaining the interaction of physical and human systems (e.g., estimate distance, calculate scale, identify dominant patterns of climate and land use, compute population density)" (Montana Office of Public Instruction, 2000). Assuming that a foundation in geographic inquiry and analysis is laid in these early and middle grades, one might expect that Montana high school students could engage in thoughtful geographic inquiry and analysis with technological tools. The stated expectations for grade 12 students in Montana are that they can, "select and apply appropriate geographic resources to analyze the interaction of physical and human systems (e.g., cultural patterns, demographics, unequal global distribution of resources) and their impact on environmental and societal changes" as well as "interpret, use, and synthesize information from various representations of the Earth (e.g., maps, globes, satellite images, geographic information systems, three-dimensional models)" (Montana Office of Public Instruction, 2000). As discussed earlier, this reference to GIS is not as detailed as other states and it is not set apart as an expectation that GIS or other technologies are used for geographic problem-solving, but the attention to vertical alignment in the Montana standards is admirable.

The Influence of National Standards

The influence of the national standards for geography and social studies is evident in the language of the state standards (National Council for the Social Studies, 1994; National Geography Education Standards Project, 1994). Many states arranged their standards to align with the categories set forth in the national standards and much of the specific terminology was identical. It is laudable that many states sought to align their standards with nationally derived expectations. It is important to recognize, though, that these national standards documents were both published in 1994. While much of the content of the standards remains relevant, the significant advances in technology since 1994 are not captured in these standards documents. Although revisions to the national standards would be one desirable course of action, the authors of state standards documents at least should be cognizant of the need to enhance and update the content of the national documents with greater attention to the emergence of geospatial technologies. Recent documents such as the NRC and UCGIS reports of 2006 as well as the National Council for the Social Studies position statement on technology in the social studies (National Council for the Social Studies, 2006) should serve as additional guides for future state standards development.

CONCLUSION

The state standards for high school geography are only one piece of the much larger puzzle of geospatial technologies in social studies education. It is important also to consider related issues such as state assessments, teacher preparation and professional development, curriculum models, instructional methods, and school technology infrastructures. We have focused on state standards because we view them as a potential disincentive to geospatial technology adoption that has not received sufficient attention in the literature. Our survey of state standards reveals that few states incorporate geospatial technologies into their standards for high school geography. This sparse attention to geospatial technologies in state standards should be viewed as an additional source of friction that has slowed the implementation of these applications by high school geography teachers. Certainly, there are inspiring stories of teachers adopting GIS and other technologies in states where they are not expected to do so (e.g., Alibrandi, 2003; Alibrandi & Sarnoff, 2006). It is also quite likely that we could point to geography classrooms in Texas and Minnesota where students are not using geospatial technologies despite the explicit expectation stated in the standards of these states. Thus, the presence of standards does not ensure implementation and a lack of standards does not ensure neglect. Yet standards are a significant force driving teacher decision-making and administrative budgeting, and the strengthening of standards related to geospatial technology is one crucial step toward realizing the promise of these technologies in K–12 education.

REFERENCES

Alaska Department of Education and Early Development. (n.d). *Alaska content standards.* Retrieved January 8, 2007, from http://www.eed.state.ak.us/contentstandards/Geography.html

Alibrandi, M. (2003). *GIS in the classroom: Using geographic information systems in social studies and environmental science.* Portsmouth, NH: Heinemann.

Alibrandi, M., & Sarnoff, H. (2006). Using GIS to answer the 'whys' of 'where' in social studies. *Social Education, 70*(3), 138–143.

Arizona Department of Education. (n.d.). *Social studies standards.* Retrieved January 8, 2007, from http://www.ade.state.az.us/standards/sstudies/articulated/

Arkansas Department of Education. (2000). *Social studies curriculum frameworks.* Retrieved January 8, 2007, from http://arkansased.org/teachers/pdf/socialstudies.pdf

Baker, T. R. (2005). Internet-based GIS mapping in support of K–12 education. *The Professional Geographer, 57*(1), 44–50.

Bednarz, S. W., & Audet, R. H. (1999). The status of GIS technology in teacher preparation programs. *Journal of Geography, 98,* 60–67.

Bednarz, S. W., & van der Schee, J. (2006). Europe and the United States: The implementation of geographic information systems in secondary education in two contexts. *Technology, Pedagogy and Education, 15*(2), 191–205.

California State Board of Education. (1998). *History-social science content standards for California public schools.* Retrieved January 8, 2007, from http://www.cde.ca.gov/re/pn/fd/documents/histsocsci-stnd.pdf

Colorado Department of Education. (1995). *Colorado model content standards for geography.* Retrieved January 8, 2007, from http://www.cde.state.co.us/cdeassess/documents/standards/geog.htm

Delaware Department of Education. (2001). *Social studies standards.* Retrieved January 8, 2007, from http://www.doe.k12.de.us/DPIServices/Desk_Ref/SSStandrev2001/9_12ssstandards.pdf

DiBiase, D., DeMers, M., Johnson, A., Kemp, K., Luck, A. T., Plew, B., et al. (2006). *Geographic information science & technology body of knowledge.* Washington, DC: American Association of Geographers and the University Consortium for Geographic Information Science.

Donaldson, D. P. (2001). With a little help from our friends: Implementing geographic information systems. *Social Education, 65*(3), 147.

Florida Department of Education. (n.d.). *Social studies standards grades 9–12.* Retrieved January 8, 2007, from http://www.firn.edu/doe/curric/prek12/pdf/socstud9.pdf

Gatrell, J. D. (2004). Making room: Integrating Geo-technologies into teacher education. *Journal of Geography, 103*, 193–198.

Georgia Department of Education. (2006). *Georgia performance standards for social studies.* Retrieved January 8, 2007, from http://www.georgiastandards.org/socialstudies.aspx

Idaho State Department of Education. (n.d.). *Geography standards for grades 6–9.* Retrieved January 8, 2007, from http://www.sde.state.id.us/dept/docs/standards/StandardsTable_SocStudies.pdf

Illinois State Board of Education. (n.d.). *Illinois learning standards for social science.* Retrieved January 8, 2007, from http://www.isbe.state.il.us/ils/social_science/pdf/goal18.pdf

Indiana Department of Education. (2001). *Indiana's academic standards for world geography.* Retrieved January 8, 2007, from http://www.doe.state.in.us/standards/HS-SocialStudies.html

Kansas State Department of Education. (2005). *Curricular standards for history-government; economics and geography.* Retrieved January 8, 2007, from http://www.ksde.org/outcomes/ssstdh.pdf

Kentucky Department of Education. (n.d.). *High school graduation requirements.* Retrieved January 8, 2007, from http://www.education.ky.gov/KDE/Instructional+Resources/High+School/Additional+Information/High+School+Graduation+Requirements.htm

Kerski, J. J. (2003). The implementation and effectiveness of geographic information systems technology and methods in secondary education. *Journal of Geography, 102*, 128–137.

Longley, P. A., Goodchild, M. F., Maguire, D. J., & Rhind, D. W. (2001). *Geographic information systems and science.* New York: Wiley.

Louisiana Department of Education. (1997). *Louisiana social studies content standards.* Retrieved January 8, 2007, from http://www.louisianaschools.net/lde/uploads/2912.pdf

Milson, A. J., DeChano, L. M., Bunch, R. L., Caito, J., & Qiu, X. (2005). GIS in K–12 education: Pedagogical tool or weapon of mass distraction? *Research in Geographic Education, 7*, 62–73.

Minnesota Department of Education. (2005). *Minnesota academic standards in social studies.* Retrieved January 8, 2007, from http://education.state.mn.us/MDE/Academic_Excellence/Academic_Standards/Social_Studies/index.html

Minnesota Department of Education. (2006). *Minnesota K–12 academic standards, required state tests, and graduation requirements.* Retrieved January 8, 2007, from http://education.state.mn.us/mdeprod/groups/Standards/documents/Instruction/001070.pdf

Mississippi Department of Education. (2004). *Mississippi social studies framework and guide.* Retrieved January 8, 2007, from http://www.mde.k12.ms.us/acad/id/curriculum/ss/frame.html

Montana Office of Public Instruction. (2000). *Montana standards for social studies.* Retrieved January 8, 2007, from http://www.opi.mt.gov/pdf/standards/ContStds-SocSt.pdf

National Council for the Social Studies. (1994). *Expectations of excellence: Curriculum standards for social studies.* Washington, DC: Author.

National Council for the Social Studies. (2006). *Technology position statement and guidelines.* Retrieved January 8, 2007, from http://www.socialstudies.org/positions/technology

National Geography Education Standards Project. (1994). *Geography for life: National geography standards, 1994.* Washington, DC: National Geographic Research & Exploration.

National Research Council. (2006). *Learning to think spatially.* Washington, DC: National Academies Press.

Nevada Department of Education. (2000). *Geography standards.* Retrieved January 8, 2007, from http://www.doe.nv.gov/standards/socialstudies/geocont.html

New Jersey Department of Education. (1999). *New Jersey social studies curriculum framework.* Retrieved January 8, 2007, from http://www.state.nj.us/njded/frameworks/socialstudies/

New Mexico State Board of Education. (2001). *New Mexico social studies content standards and benchmarks.* Retrieved January 8, 2007, from http://www.nmlites.org/downloads/standards/stand_ss.pdf

New York State Education Department. (1996). *Learning standards for social studies.* Retrieved January 8, 2007, from http://www.emsc.nysed.gov/ciai/socst/pub/sslearn.pdf

North Carolina Department of Public Instruction. (2003). *Social studies standard course of study.* Retrieved January 8, 2007, from http://www.ncpublicschools.org/curriculum/socialstudies/scos/

Ohio Department of Education. (2002). *Social studies academic content standards.* Retrieved January 8, 2007, from http://www.ode.state.oh.us/

Oklahoma State Department of Education. (2002). *Priority academic student skills: High school social studies.* Retrieved January 23, 2007, from http://sde.state.ok.us/home/defaultie.html

Pennsylvania Department of Education. (2002). *Academic standards for geography.* Retrieved January 8, 2007, from http://www.pde.state.pa.us/stateboard_ed/lib/stateboard_ed/pdfgeography.pdf

Radinsky, J. (2008). GIS for history: A GIS learning environment to teach historical reasoning. In A.J. Milson & M. Alibrandi (Eds.), *Digital Geography: Geospatial technologies in the social studies classroom* (pp. 99–117). Charlotte, NC: Information Age.

South Carolina Department of Education. (2005). *South Carolina social studies academic standards.* Retrieved January 8, 2007, from http://ed.sc.gov/agency/offices/cso/standards/ss/documents/9INEZSocialStudiesStandards.pdf

South Dakota Department of Education. (2005). *Graduation requirements.* Retrieved January 8, 2007, from http://doe.sd.gov/ofm/prf/docs/2006GraduationRequirementsDirections.pdf

South Dakota Department of Education. (2006). *Social studies content standards.* Retrieved January 8, 2007, from http://doe.sd.gov/contentstandards/social/docs/Full_Social%20Studies.pdf

Tennessee Department of Education. (2001). *Social studies curriculum standards.* Retrieved January 8, 2007, from http://www.state.tn.us/education/ci/cistandards2001/ss/cissworldgeography912.htm

Texas Education Agency. (1998a). *Texas Essential Knowledge and Skills for Social Studies.* Retrieved January 8, 2007, from http://www.tea.state.tx.us/rules/tac/chapter113/ch113c.html#113.34

Texas Education Agency. (1998b). *Texas state graduation requirements for students entering grade 9 in 1998–99 and thereafter.* Retrieved January 8, 2007, from http://www.tea.state.tx.us/teks/handbook/7Ch74Gr9in9899.PDF

Utah State Office of Education. (2002). *Secondary core curriculum: Social studies 7–12.* Retrieved January 8, 2007, from http://www.schools.utah.gov/curr/core/corepdf/SoSt7-12.pdf

Washington Office of Superintendent of Public Instruction. (2003). *Geography essential academic learning requirements.* Retrieved January 8, 2007, from http://www.k12.wa.us/curriculumInstruct/SocStudies/geographyEALRs.aspx

CHAPTER 3

TOWARDS AN INTERNATIONAL GEOSPATIAL EDUCATION COMMUNITY

Joseph J. Kerski
Environmental Systems Research Institute

At the dawn of the 1990s, amid the flurry of early development of Geographic Information Systems (GIS), one of its primary architects, Dr. Michael Goodchild (1992), paused and reflected on the situation to which GIS had evolved. Examining developments that had occurred during the previous 25 years, he claimed that GIS was no longer just a tool, but it represented a way of thinking about problems from a spatial perspective. He asserted that given the fact that GIS had its own research agenda, development initiatives, and an active community of users, GIS had evolved into Geographic Information *Science*. While it had strong ties to geography, computer science, cartography, geodesy, remote sensing, and other disciplines, Geographic Information Science was more than just a sum of the parts—it had become its own distinct discipline.

Effective teaching with GIS had to wait for hardware, software, accessibility and availability of data, computer and human networks, educational standards, awareness, and other forces to mature to the point to where it became technologically and pedagogically feasible to use in the classroom. It wasn't until Goodchild claimed the arrival of Geographic Information Science when teaching with GIS first appeared in scattered classrooms at

Digital Geography: Geospatial Technologies in the Social Studies Classroom, pages 61–74
Copyright © 2008 by Information Age Publishing
All rights of reproduction in any form reserved.

the secondary and university level. However, advances since that time have brought us to the point whereby in 2007, I believe that GIS education has evolved beyond the enthusiastic dedication of isolated instructors to the point of being an international education community.

As the 1990s progressed, GIS software functions increased in number and sophistication, and migrated from command-line input to graphical user interfaces. Processing speed, the ability to manipulate large databases, and the database models themselves improved manyfold. Spatial data became more widely available and in a greater range of themes and scales.

Since 2000, GIS has increasingly integrated Internet-based data with software residing on a desktop computer. At the same time, web-mapping tools have become increasingly sophisticated, to the point of allowing the user to run spatial and attribute queries and to upload their own data. Even so, GIS is more than hardware, software, procedures, and spatial data. GIS is a network of people who collaborate to solve integration and technical problems and to move forward. Some have argued (Bijker, Hughes, & Pinch, 1987; Pannabecker, 1991) that technology is socially constructed. In this context, the "relevant social groups"—those who actively contribute to the construction of GIS as a viable technology for use in educational contexts—are a diverse but cohesive group who are committed to advancing GIS in education and society. The convergence of web-based mapping, virtual globes, GIS, remote sensing, and GPS technologies is often referred to as *geospatial technologies*. Thus, the education community that has adopted and embraced these technologies and methods can be referred to as the *International Geospatial Education Community*.

The Characteristics of the International Geospatial Education Community

What makes a community? Community comes from two Latin words meaning "with gifts," and therefore, suggests a sense of altruism, benefit, and learning from each other that comes from working together. Whether in the physical world, online, (Goodfellow, 2005), or some combination of the two; a community is held together by some common identity or interests (Parse, 2003). A shared interest, or intent, gives members of the community a reason to build relationships and affiliations out of which they can learn from one another and make an impact on the society around them. Communities help generate a shared language, ways of working, a collective memory, and shared experience of those that join the group.

I argue that the above traits characterize the geospatial education community. This community embraces the four components of what Wenger (1998) claims are necessary to characterize social participation as a process of learning and of knowing. These include Meaning: A way of talking about our ability to experience our life and the world as meaningful; Practice: A way of talking about the shared resources, frameworks, and perspectives that can sustain mutual engagement in action; Community: A way of talking about the social configurations in which our enterprises are defined as worth pursuing and our participation is recognizable as competence; and Identity: A way of talking about how learning changes who we are and creates personal histories in the context of our community.

The community is young but amazingly cohesive, particularly in its mission and agenda. Its mission is to further the spatial perspective through the use of geotechnologies in education and throughout society. Its agenda is varied, but usually encompasses one or more of five components.

First, similar to other educational communities (Kadel, 2005), this community has struggled not only with the lack of research in this field, but also what its research agenda should be. Second, the community seeks to further research the effectiveness and implementation of geospatial tools and methods in formal and informal education. This includes the development of assessment instruments that can truly measure the effectiveness and implementation of GIS in the affective and content domains. The reason is that the community realizes that for policymakers and education administrators to take notice of the value of spatial thinking and learning, the research community must provide the hard evidence that investing in geotechnology-based, inquiry-driven methods and tools is worth the effort.

A related research goal is to uncover challenges to the implementation of geotechnologies so that the challenges can be mitigated or removed. While the community seeks to promote the use of geotechnologies and spatial thinking in teaching and learning, it also recognizes that it is difficult to "prove" their effectiveness, particularly when "effective" often means according to scores on standardized tests that place so much emphasis on memorization of facts. Community participants have a balanced view, despite their enthusiasm for the subject, that these technologies and methods will not solve all that ails educational systems around the world: Many are veteran teachers who have seen technologies come and go, many of which promised much but delivered little, and many of which were adopted with little concern for their efficacy or importance (Stevens, 2004). They are keen to investigate challenges to effective implementation in the curriculum, albeit with an ultimate goal to reduce or eliminate these challenges. Thus, the research is aligned with action-oriented geography research, or research aimed at changing social thinking and policy, prominent during the 1960s but enjoying a resurgence from the late 1990s (Pain, 2003).

Third, the community has very broad geographic and educational aims. It seeks to teach with geotechnologies in broader circles that encompass settings in formal and informal education, urban and rural environments, and in an expanding array of disciplines, and in more effective ways, all over the world. Fourth, the community seeks to develop curriculum to support teaching with geotechnologies. Community leaders are earnestly constructing lessons, scope and sequence models, and entire books providing educators with data, assessments, and lessons that they can use in geography, physics, chemistry, environmental studies, mathematics, history, sociology, earth science, environmental studies, agriculture, and other disciplines. Fifth, the community continually seeks to communicate the message that increased use of geotechnologies and spatial thinking methods in education will result in more informed decisions at the local, regional, national, and international scales, and will result in improvements in quality of life for people and health for the planet. They believe that the message needs to be carried far beyond the walls of the educational institution, to local education authorities, to the news media, to scientists, and to national policymakers.

In the days before the community became established, individual educators and those who supported them sought to further this agenda alone. They represented the "individual change agents" that Rogers (1995) describes as an indicator of early diffusion of innovations. The community's presence has emphasized and hastened the role of partnerships in accomplishing its mission. The community seeks to form partnerships among primary and secondary schools, vocational-technical schools, community colleges, and universities.

Perhaps the community's characteristic that shows the most maturation is embodied by these active partnerships that move the agenda of the community forward. Partnerships result in increased training and teaching opportunities, curriculum, and research. A partnership between NASA and the National Council for Geographic Education (NCGE), for example, generated new remote-sensing and GIS-based curriculum at the Grade 5 level. The My Community Our Earth (http://www.aag.org/sustainable) initiative through the Association of American Geographers encourages students around the world to examine community issues through geographic thinking and mapping tools. Continent-wide programs such as HERODOT (The Thematic Network for Geography Teaching and Training at http://www.herodot.net/), country-wide initiatives in Denmark (http://www.gogis.dk/) and New Zealand (http://www.gismaped.co.nz/), or the statewide GIS initiatives in Virginia and Utah and elsewhere in the USA are the result of successful partnerships among nonprofits, industry, higher education, primary and secondary schools, and government agencies. The international geospatial education community shares a language that includes a combination of terms in education, GIS, and geography. The community represents diverse backgrounds and educational levels. Their job titles include those who are in full-time education, such as schoolteachers, professors, and educational consultants, to those who support education through their role in government, academia, nonprofit organization, or private company. The community includes those who teach *about* GIS and those who teach *with* GIS (Sui, 1995). The community of those who teach *about* GIS began primarily in community colleges and universities, and in nonformal settings such as National 4-H, but now is finding its way into secondary and vocational-technical schools. The community of those who teach *with* GIS has a core in secondary education, but has spread to primary education and to universities.

One is struck by the high level of professional engagement and activity of many in this community. They frequent the same conferences, such as those of an educational nature—such as the American Educational Research Association, the National Science Teachers Association, the National Council for Geography Education, the Geographical Association, those focused on geography and science—such as the Association of American Geographers, the Geological Society of America, and those more focused on GIS—such as regional GIS conferences, the Urban and Regional Information Systems Association, the Geospatial Information Technology Association, and the Environmental Systems Research Institute's User Conference, They often engage in joint research, scholarship, and teaching. In what may be the most profound indicator of community, one hears an amazing number of stories about community members spending personal vacation time together.

The tenets shared by GIS educators have been remarkably constant over the past two decades and the community has developed around these tenets. The first commonly held belief is that the purpose of using geotechnologies, whether on the desktop computer or on the Internet, is not to make the perfect map. The goal is to foster inquiry and analysis about the Earth and its people. Furthermore, the tools themselves are not the end goal, but rather, they are the *means* to the end goal of enhanced ability to think about problems from a spatial framework. Educational standards and national curricula are used by this community as justification for the higher order thinking skills, rather than teaching *to* the standards. By uncovering the inquiry basis underlying each standard, they find that a strong case can be made for spatial analysis. Similar to what occurs with the use of other inquiry-based methods, not everyone in the classroom will learn the same thing when they use GIS, which is difficult to manage in the classroom. However, many educators find that what students *do* learn is valuable. They are learning not just about natural hazards or demographics, but are rethinking their role in society, rethinking lifelong learning, and, often, rethinking their own career path. While most geotechnology is still not plug and play, and may never be so due to the inherent complexities of geographic problem-solving, many educators find that one does not have to be an expert to use some of the tools for the classroom. Most students will be interested in GIS on the desktop or on the web, often require little "coaching," and often exhibit a high level of peer mentoring. However, the instructor's role is still critical. It is the instructor who frames the questions that engage higher order thinking skills and in a spatial context, to prevent the exercises from becoming simply a random browse through maps. Because of this, the instructor should not, in the view of community members, spend most of his or her time mastering GIS skills or downloading and formatting spatial data. The student engagement ensures that most students will master the technology faster than the teacher will. Therefore, the use of geotechnologies in one sense free the instructor to do what he or she is most skilled at doing and what he or she enjoys doing the most—to frame the inquiry-based questions.

While it is true that most of the community leaders have an affinity for geography and education, what drives the community forward is a strong belief in a shared mission. This mission is not simply to conduct GIS-based research, curriculum development, or teaching, but the community seeks to increase the spatial literacy of the entire global society. In addition, community members share a sense of urgency that spatial literacy and skills need to be acquired by society as soon as possible, if the society is going to make it to the other side of the 21st century in a sustainable, healthy, and growing way.

Because the international geospatial education community is just beginning, those who could be considered its founders remain some of its most active. The technology has changed, and the community now embraces technology beyond desktop GIS; from the educational applications of remote sensing to web-mapping, virtual globes, and Global Positioning Systems. Far from stagnating, the community adds new leaders each year and by 2007 may number over 2000 active participants. The primary forum for meeting may well be the annual ESRI Education User Conferences, which grew from 100 people in 2000 to over 800 by 2006.

The Development of the International Geospatial Education Community

Why and how did the international geospatial education community develop? The reason why the community developed is rooted in the fact that the implementation of a new set of tools and methods in any educational environment is highly dependent on the interaction of people. People network to share everything from pedagogical implementation strategies to technical tips. Increasing the need for this network is the fact that GIS is a complex set of tools, procedures, and an inquiry-based approach that is completely unlike the predefined digital atlases such as Encarta that had existed previously.

In terms of *how* the international geospatial education community developed, it was in part dependent on technological and educational forces, but also on active community members who pushed its agenda forward. Advances in computing technologies ensured that GIS progressed beyond the minicomputers and mainframes of government and higher education to become accessible to teachers on a personal computer. Until 1990, nearly every computer workstation running GIS included an expensive graphics monitor, was attached to a digitizing tablet, and required a minicomputer or mainframe computer for data storage. As GIS became implemental in a standard computer lab, school computer lab managers, teachers and computer support staff began running the software in school computer laboratories, and on the increasing number of computers appearing within the classrooms themselves. Because the manipulation of digital maps, grids, tables, and images are complex operations, the increase in computing power, memory, and storage space were key to the expansion of GIS in education just as they were key in its expansion in natural resources, planning, energy, and other industries.

Software that was more powerful and easier to use was a critically important innovation for educators who did not have the time to develop expertise in GIS. Two key software advancements were the change from command-line to menu-driven GIS software, followed by the GIS emulating standard Windows and Macintosh software interfaces of the day. Once GIS software had the same look-and-feel as other software, teachers could transfer skills they had learned on word processing, graphics, and database programs to GIS software.

Because GIS is absolutely dependent on the use of spatial data, the increasing availability of data sets at first from federal government agencies, followed by state and local governments, nonprofit organizations, private industry, and higher education provided enormous utility to everyone using GIS. These data sets included aerial photography, satellite imagery, elevation, hydrography, land use, transportation, biomes, geology, population, and others collected at local to global scales. Perhaps the availability of data provided a larger boost in education than in some other sectors of society because teachers and students lacked the hardware and computer networks necessary to generate their own base information and had to rely on others for it. Finally, it was important that government data that arose from National Spatial Data Infrastructure (NSDI) initiatives in the USA and portals from non-government agencies provide free and low-cost data, because for the most part, educators had no funds or budget to support the purchase of spatial data.

Coupled with the increased amount of data was the expansion of the means by which educators could access spatial information. No longer confined to 9-track tape and other media that few educators could work with, educators and their students could access data at first through File Transfer Protocol (FTP), and later through HTML-coded web pages through the World Wide Web. The next development was the advent of mapping portals such as the Federal Geographic Data Committee's Geospatial One-Stop, from which the user could download spatial data. Because spatial data tends to be large in size, requiring much bandwidth, the gradual disappearance of the use of telephone lines for Internet access served to advance the use of GIS in schools. Within schools, wired and then wireless networks made possible the transfer of large spatial data files that are so fundamental to digital mapping success.

In order for a complex toolkit such as GIS to become adopted by the schools, teachers had to receive support. Initiatives such as enhancement and support for geography education in the USA from National Geographic through the development of geographic alliances and teacher training (Salter, 1987) represent a targeted, coordinated, national program that was maintained for over 15 years. However, GIS training in the USA was largely conducted by a private software vendor and through a train-the-trainer model by teachers who had themselves attended a hands-on, face-to-face GIS-based training institute that ranged in length from a few days to a few weeks. To be sure, some geographic alliances and professional societies occasionally sponsored a training institute, but private companies and consultants led the way, most notably, Environmental Systems Research Institute (ESRI). ESRI has led hundreds of training events over the past 15 years, including two at Texas State University during the late 1990s where many teachers who in turn trained other teachers came together. From these institutes, trained teachers began conducting their own training events, such as those by GISetc, Compass, and WorldViews, which are educational technology companies established by educators.

More recently, online courses and certificate programs (such as those at the University of Montana and Eastern Michigan University) have also trained hundreds of teachers. Unlike GIS workshops for GIS professionals, these online and face-to-face courses spend as much time discussing pedagogical methods of geotechnology inclusion as they do on GIS methods and tools. Even the GIS immersion emphasizes the investigation of a geographic issue such as epidemics or hurricanes, rather than GIS skills. This distinctive feature of GIS training for educators is one key reason why educators adopted the technology—in the words of the diffusion of technology and the educational community—the training used their common shared language as educators. By contrast, most educators trained alongside a city planner or a utility engineer in a standard GIS did not adopt the technology. After a series of GIS in education conferences during the 1990s, an educational GIS conference was established by ESRI in 2000 and now attracts over 800 educators annually from all over the globe.

In part because of the aforementioned changes, the number of subject disciplines in which GIS is used has expanded to new subject disciplines and has grown within existing ones. GIS in education in the USA began around 1990 through the efforts of pioneering geography and science teachers. More science teachers than geography teachers were using GIS because of more frequent and sustained computer-based professional development, better access to computers, and more familiarity and comfort with inquiry-based methods of instruction than their geography colleagues. By the time a decade had passed, Kerski (2003) still found more science teachers using GIS than geography teachers in a survey of over 1,500 teachers who owned GIS software. However, since 2000, GIS has expanded beyond geography into other subjects in the social studies, beginning with history, and continuing into economics and civics. In the sciences, the use of GIS extended beyond chemistry and earth science into environmental studies, physics, and biology. GIS can also now be found in classes in mathematics, language arts, technology, and business. Indeed, so many success stories existed that they were compiled into a popular book entitled *GIS in Schools* (2000).

Following recent initiatives stemming from U.S. Department of Labor concerns over the lack of trained GIS professionals in society (Gewin, 2004), GIS has expanded into classes in technology and workforce development. In addition, GIS has expanded beyond the walls of the school into informal educational settings. These include local youth development programs, such as HopeWorks 'N Camden (Kelly, 2005), whose mission is to enhance the lives of inner city Camden youth by expanding the learning opportunities available to them, specifically focusing on technology training. The training includes website design, GIS, computer networking, computer repair, and video. The program seeks to reduce the high school dropout rate specifically among African-American and Hispanic American youth in Camden, and to create jobs, business development, and educational opportunities there. They also extend to national efforts, such as the National 4-H GPS-GIS initiative (Tate, 2005). Supported by the USDA and ESRI, this program includes GIS software, training, and supplemental materials, supporting 4-H Clubs new to GIS and GPS technology and also for clubs in which students are already examining issues ranging from landfills to sustainable agriculture in their communities. American Forests, the nation's oldest nonprofit citizens' conservation organization, has joined with ESRI on the 4-H grants by providing its CITYgreen extension for ArcView GIS software for students doing tree inventories as part of their community service project.

Despite a slow start, research efforts in GIS in education are gaining strength in numbers and effectiveness. Because GIS is not a subject in most primary and secondary schools, school officials need to be convinced of its value and its curricular fit, and they require evidence from educational research in order to justify support for the teachers who advocate its use. Early GIS research attempted to move beyond the largely anecdotal efforts of individual success stories in individual classrooms that was prevalent in the literature through the 1990s to such topics as student's attitudes and achievement (Baker & Case, 2003), and the implementation at the national level and its effectiveness in individual schools (Kerski, 2003). More recently, the GIS education research community received support for its theoretical base from a study by the National Research Council, National Academy of Sciences, *Learning to Think Spatially: The Incorporation of Geographic Information Science Across the K–12 Curriculum* (National Research Council, 2006). This report emphasized the critical need for developing spatial concepts and skills beginning very early in the primary curriculum and continuing past university level.

The educational standards movement across the world has served to advance the use of GIS in the classroom. While concerns have been voiced about the overemphasis on standardized tests that are based on rote memorization that may stifle inquiry-based teaching methods and technologies such as GIS, most educational content standards and national curricula focus on the higher-order thinking skills such as analysis, synthesis, and evaluation over knowledge and comprehension. These are the more advanced sections of Bloom's Taxonomy (1956), and are recognized as essential for students to compete in today's globalized economy. GIS is not just a technology, but includes spatial analysis methods that were created to do just that—*inquire* about the planet, is people, and its patterns, trends, and relationships. Indeed, geotechnology education has evolved to the point where it influences regional, state, and national standards, beginning with the inclusion of geographic tools in *Geography for Life* (Geography Education Standards Project, 1994).

Educational drivers that promoted the use of GIS in schools included instruction that adhered to the tenets of constructivism, that provided inquiry-driven, problem-based learning, and that assessed student work in an authentic way that emulated the workplace and whole learning. This was instruction that emphasized multidisciplinary education, team-based collaboration, peer mentoring, and that was focused on authentic projects. Good instruction was viewed as that which provided job skills and emphasized using real-world technologies, data, and contexts, as well as enhanced critical thinking, problem-solving, and decision-making skills (Alibrandi, 2003).

Concern has heightened over recent years about the reduction in field work in Earth Science and Geography education, but also outside the educational system, with young people spending more and more time in front of the computer, and less time outside (Louv, 2005). Coupled with this are increasing administrative complications in transporting and obtaining permission to get students to field locations, GIS with GPS are viewed as an excellent means of motivating students to study local issues and to apply those skills to regional, national, and international problems. Geotechnologies are also considered to be an excellent bridge between schools and their communities, not just because the same technology is used by local governmental and community planners, but because the very issues that communities are grappling with can be investigated by students as well (Orton Family Foundation, 2003). Educators increasingly value geotechnologies as part of essential citizenship education. Students who study local issues while still in primary and secondary school may be more active in their community as adults and are better poised to conduct research at university level (Kerski, 2004).

At the same time that this small but very active group of educators and researchers was studying and implementing GIS in education, easy-to-use geographic technologies began to be offered to the general public for everyday work and play. Over a very short time, Global Positioning Systems (GPS) receivers and online map services became in demand not just by the geographic community, but also by a much broader audience. Affordable GPS receivers, aided by increased accuracy when the Selective Availability code was turned off in 2000, online mapping services such as MapQuest and Yahoo maps, and GPS-enabled cell phones rapidly became a common part of life. Because educators, parents, and students are surrounded by geotechnologies, the awareness of these services is at an all-time high. Coupled with awareness is the interest level, which soared even higher with the advent of 3-D Virtual Globe technologies such as ArcGIS Explorer, NASA WorldWind, and Google Earth. Educators are certainly less cognizant about what these services offer education than what they offer recreation, navigation, employment, and research. However, the first step in Rogers' (1995) diffusion of innovation—awareness—has been largely met through societal inundation by online mapping, virtual globes, GIS, and GPS.

Public awareness of environmental issues may also be at an all-time high. Sustainable development, climate change, urban sprawl, water quality and availability, crime, energy, migration, and biodiversity are only a few of the terms on the public's mind and in daily newspapers. These are not only relevant, but many carry with them the sense of urgency—that they are complex problems that need to be solved, and will not be solved unless we are regularly engaging young people in thinking about them. While the public still has not made the connection between these issues and the applicability of geotechnologies in education, or even the fact that the discipline of geography regularly grapples with these issues, parental, teacher, and public awareness of these issues is encouraging many educators to increasingly teach them in the classroom.

Despite all of the aforementioned advances, GIS would have remained a technology and workforce skill in education, confined to those teaching *about* GIS versus teaching *with* GIS (Sui 1995), without simultaneous advancements in curriculum. Curricular units are different from technical training modules, and are essential for educators who seek to teach with GIS because they provide the means by which *concepts* are taught, not technical skills. Certainly, technical skills are learned through these units, but their primary emphasis is to teach about a specific topic and process, such as population migration or the relationship between climate, land cover, and human impact on the landscape through land use. In short, they teach problem solving, and have a particular focus on the "whys of where"—geographic inquiry. Curriculum began with lessons that individual teachers created in the classroom and progressed to these same lessons uploaded to the Internet, and expanded, modified, and shared with other educators. It then extended to lessons, assessments, and extensions in books such as *Mapping Our World* (Malone, Palmer, & Voigt, 2002) and *Community Geography* (Zanelli-English & Feaster, 2003), Internet-based libraries such as ArcLessons (http://www.esri.com/arclessons) and national efforts such as New Zealand's active GIS education community (Olsen, 2002) and Denmark country-wide GIS in education program (Jensen, 2004), to international initiatives such as the geotechnology focus in the Thematic Network for Geography teachers and lecturers in Higher Education (HERODOT) (Donert, 2005).

Other forces not mentioned here have fostered the creation of the international geospatial education community. At the same time, forces old and new continue to ebb and flow, acting as challenges to the community's effectiveness and indeed, the implementation of geotechnologies in education. The community has a big job to do, but the progress thus far will likely encourage the community to press on toward its goals.

REFERENCES

Alibrandi, M. (2003). GIS *in the classroom: Using geographic information systems in social studies and environmental science.* Portsmouth, NH: Heinemann.

Baker, T., & Case, S. (2003). The effects of GIS on students' attitudes, self-efficacy, and achievement in middle school science classrooms. *Journal of Geography, 102*(6), 243–254.

Bijker, W., Hughes, T., & Pinch, T. (1987). *The social construction of technological systems.* Cambridge, MA: MIT Press.

Bloom, B. (1956). *Taxonomy of educational objectives, handbook 1.* New York: Longman.

Donert, K. (2005). *Higher education GIS in geography: A European perspective.* Liverpool, UK: Liverpool Hope University.

Geography Education Standards Project. (1994). *Geography for life: National geography standards.* Washington, DC: National Geographic Research and Exploration.

Gewin, V. (2004). Mapping opportunities. *Nature, 427,* 376–377.

Goodchild, M. (1992). Geographic information science. *International Journal of Geographical Information Systems,* 6(1), 31–46.

Goodfellow, R. (2005). Virtuality and the shaping of educational communities. *Education, Communications & Information, 5*(2), 113–129.

Jensen, T. (2004). GIS in environmental and wind energy- Danish high schools. *2004 ESRI Education User Conference Proceedings.*

Kadel, R. (2005). Putting the pieces together: Exploring the idea and purpose of developing a research agenda for educational technology. *Learning and Leading With Technology, 32*(8), 28–29.

Kelly, M. C. (2005, August). Non-profits and volunteers on the leading edge of GIS? *Directions Magazine.* Retrieved June 12, 2007, from http://www.directionsmag.com/article.php?article_id=923&trv=1

Kerski, J. (2003). The implementation and effectiveness of GIS technology and methods in secondary education. *Journal of Geography, 102*(3), 128–137.

Kerski, J. (2004). GIS for citizenship education. In A. Kent & A. Powell (Eds.), *Geography and citizenship education: Research perspectives* (pp. 129–138). London: University of London Institute of Education for the International Geographical Union Commission on Geographical Education.

Louv, R. (2005). *Last child in the woods: Saving our children from nature-deficit disorder.* Chapel Hill, NC: Algonquin Books.

Malone, L., Palmer, A., & Voigt, C. (2002). *Mapping our world: GIS lessons for educators.* Redlands, CA: ESRI Press.

National Research Council. (2006). *Learning to think spatially.* Washington, DC: National Academies Press.

Olsen, A. (2002). Using GIS software in school teaching programmes: An initial survey. *New Zealand Journal of Geography, 113,* 17–19.

Orton Family Foundation. (2003). *Making community connections: The Orton Family Foundation Community Mapping Program.* Redlands, CA: ESRI Press.

Pain, R. (2003). Social geography: On action-orientated research. *Progress in Human Geography, 27*(5), 649–657.

Pannabecker, J. (1991). Technological impacts and determinism in technology education: Alternate metaphors from social constructivism. *Journal of Technology Education, 3*(1), 43–54.

Parse, R. (2003). *Community: A human becoming perspective.* Boston: Jones and Bartlett Publishers.

Rogers, E. (1995). *The diffusion of innovations.* (4th ed.). New York: Free Press.

Salter, C. (1987). The nature and potential of geographic alliance. *Journal of Geography 86*(5), 211–215.

Stevens, R. (2004). Why do educational innovations come and go? What do we know? What can we do? *Teaching and Teacher Education 20*(4), 389–396.

Sui, D. (1995). A pedagogic framework to link GIS to the intellectual core of geography. *Journal of Geography 94*(6), 578–591.

Tate, T. (2005). Expanding the horizons of youth and communities: 4H GIS/GIS technology teams. *2005 ESRI User Conference Proceedings.*

Wenger, E. (1998). *Communities of practice: Learning, meaning, and identity.* Cambridge, UK: Cambridge University Press.

Zanelli-English, K., & Feaster, L. (2003). *Community geography: GIS in Action.* Redlands, CA: ESRI Press.

PART II

SOCIAL STUDIES EXPEDITIONS WITH GEOSPATIAL TECHNOLOGIES

CHAPTER 4

BEYOND INTERACTIVE MAPPING

Bringing Data Analysis with GIS into the Social Studies Classroom

Daniel C. Edelson
National Geographic Society

David A. Smith
GLOBE

Matthew Brown
Inquirium, LLC

For more than a decade, educators have recognized the promise of geographic information systems (GIS) as a technology to support learning through inquiry across the social and natural sciences (e.g., TERC, 1995). The history of GIS in education can be viewed from both a "glass is half full" and a "glass is half empty" perspective. The half-full story is that rapidly increasing numbers of teachers have received training on GIS, many lesson plans and books on using GIS in the classroom have been published, and large numbers of students have had opportunities to use GIS in the context of social studies and science classrooms. As with any educational innovation, though, there is a half-empty story. For GIS, that story

Digital Geography: Geospatial Technologies in the Social Studies Classroom, pages 77–98
Copyright © 2008 by Information Age Publishing
All rights of reproduction in any form reserved.

includes the fact that the overall percentages of teachers and students with GIS experience is very small, that the percentage of teachers who have received training and then go on to implement in their classroom remains small, and that the visions of students using GIS to engage in something approaching authentic inquiry is only rarely realized.

Recognizing that the half-full story provides ample reason for optimism regarding the promise of GIS as an agent for educational improvement, the Geographic Data in Education (GEODE) Initiative at Northwestern University has been engaged in a long-term research and development effort to address the challenges exposed by the half-empty story. As part of its R&D efforts, the GEODE Initiative is engaged in research on the development of GIS tools that are better suited to the needs of educators and students and to the constraints of educational contexts than the professional GIS tools that were designed for very different audiences and contexts of use. There are many reasons that GIS has achieved neither the broad use nor the kind of use that many have envisioned for it, and there have been many attempts to address these reasons through classroom research, professional development, and curriculum development. For most of the efforts to integrate GIS into education, the GIS tools themselves have been largely immutable. Educators have only had access to GIS tools that were already in the commercial market, despite having been developed for other users and contexts. The customization of GIS tools for educational use has been limited by the software provided by vendors for educational purposes and by vendors' reluctance to put too much functionality in a low-cost alternative to their tools for GIS professionals.

As a learning technologies research group, however, the GEODE Initiative has engaged in research on the design of GIS software itself by creating its own GIS tools. Over the course of a 13-year research program, we have been studying the challenges to use of GIS software by students and teachers (whom we characterize as novice and occasional users) and designing new GIS tools that reduce those challenges. Our goal is to bring the power of GIS tools to educators and students through tools that provide the structure and support to make them usable to them within the time and resource constraints of real world educational contexts.

As part of this research program, we have devoted considerable effort to analyzing the obstacles to the widespread use of GIS to support inquiry in classrooms, with a primary focus on middle and high school. One of the things that we have observed is that there is a substantial increase in difficulty between the interactive mapping functionality in GIS tools and the data analysis functionality. That is, novice and occasional users of GIS are able to use the tools to manipulate maps with minimal training. However, the use of GIS tools to conduct either quantitative analyses or database queries requires users to both develop conceptual models of GIS data and functionality, and learn specific procedures with software. We have seen a concomitant drop off between the number of classrooms that use GIS for visualization and those that use it for data analysis. Since the level of inquiry that can be achieved in a GIS without engaging in data analysis is severely limited, this threshold is a major obstacle to achieving the vision of widespread use of GIS for inquiry-based learning.

A central focus of our software research and development efforts has been understanding and reducing the challenges to data analysis with GIS. In this chapter, we describe some of the insights we have reached regarding these challenges and some of the strategies we have developed to address them. We do so through an illustrative case that presents our vision of GIS use to support meaningful inquiry in a high school history classroom. We then use that case to organize our discussion of challenges and design strategies. Throughout, we describe My World GIS™, the geographic information system we have developed both as a platform for research and as a mechanism for making the lessons of that research available to the educational community.

AN INVESTIGATION OF HISTORICAL CENSUS DATA

In this section, we present a vignette describing an investigation of GIS data by two high school history students using the data assembled by the *GIS for History* project (Radinsky, this volume). We constructed this vignette to illustrate our vision for how ordinary high school students, with appropriate guidance, can use a GIS tool to conduct meaningful investigations. Although the details of this vignette are fictitious, we have based it on more than a dozen years of experience with GIS and other data analysis activities in schools (e.g., Edelson, Gordin & Pea, 1999; Loh et al., 2001; Kyza, 2004)

Background

Ms. Rivera is a high school social studies teacher with 10 years of experience. She has recently completed a master's degree in advanced teaching at a local university that included coursework on the use of GIS tools to support quantitative data analysis in history and geography education. Based on that coursework, she has re-designed her U.S. history course to include a three-week open-ended research project, consisting of two phases. In the first phase, students use geospatial data to investigate a hypothesis about a historical event. Once they have identified a pattern in the data that confirms or contradicts the hypothesis, they engage in traditional historical research with primary and secondary sources to construct a narrative, causal explanation for the pattern they have identified. To prepare students for this project, Ms. Rivera devotes time throughout the school year to developing the inquiry skills to support both phases of the project. In this vignette, we only present the first (data analysis) phase of the research project, for which Ms. Rivera has prepared the students by teaching them basic geospatial data visualization and analysis skills through repeated experiences over the course of the year. Ms. Rivera structures students' investigations by requiring them to keep a running record of their activities and findings and review the status of their investigation with her at regular intervals. She also circulates throughout the classroom during the time that students are engaged in their research project, asking and answering questions as she goes.

Latonia and Gary, two of Ms. Rivera's students, have chosen to investigate the "Great Migration," which is the name given by historians to a major movement of African Americans away from the rural South during the early twentieth century. They had studied the Great Migration earlier in the year and had had the opportunity to work with a large collection of data that included African-American population, by county, at decade intervals from 1879 through 1960. They had learned, at that time, that the trend of migration was from the rural South to the North and to cities. After some discussion, Latonia and Gary proposed the following hypothesis to Ms. Rivera for their investigation:

> As communication and transportation improved after 1900, the rate of migration from south to north and from rural to urban communities increased.

Ms. Rivera complimented them on their hypothesis as the basis for this assignment because it describes both a trend that can be explored quantitatively with their census data and a cause that can be explored qualitatively using narrative historical accounts. Before allowing them to begin their investigation, Ms. Rivera asked them to formulate a plan for testing their hypothesis with the data and a description of what pattern they would expect to find in the data to support or reject their hypothesis. They decided to use the period from 1900 to 1960 as their study period and to compare the rate of population change before and after 1930. They predicted that in the rural south, the loss of population would be higher in the later period, as would the gain in population in the north and urban south.

Investigation Day 1

With this plan and prediction approved by Ms. Rivera, they set out to calculate the change in population during the periods from 1900 to 1930 and 1930 to 1960. First, they created two new data records for each county by sub-

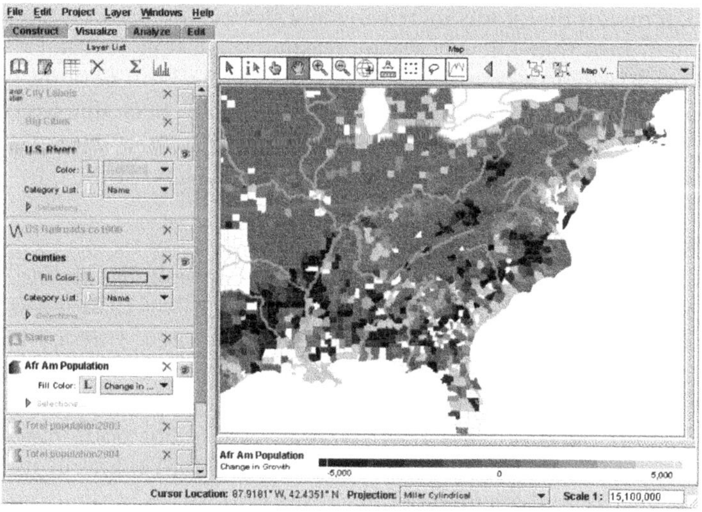

Figure 4.1. A map showing the "change in growth" of African American Population between 1900–1930 and 1930–1960. In this gray scale map, and those that follow, blue appears as lighter shades of gray and red as darker. White regions indicate missing data.

tracting the 1900 African American Population from the 1930 African-American population for each county ("1900–1930 growth"), and the 1930 population from the 1960 population ("1930–1960 growth"). They then created a third new record for each county by subtracting the "1900–1930 change" from the "1930–1960 change." They titled this third data value "change in growth." When they looked at the result (Figure 4.1), instead of seeing a pattern in the data that distinguished the rural South from the North and urban South, they found themselves looking at a patchwork quilt.

After verifying that they had performed the computation that they had intended, they called Ms. Rivera over for help making sense of their map. She asked them what they would expect to see if their hypothesis were correct. They answered quickly that they had expected to see fields of blue (indicating an increase in population *growth*) in the North and in urban areas in the South. Similarly, they expected the rural South to be a field of red (indicating an increase in population *loss*). When Ms. Rivera asked if they could see the pattern they were looking for anywhere, they realized they were able to see the increase they expected in the Upper Midwest and the Northeast, but not in the patchwork of reds and blues elsewhere. She then asked them what they thought it meant when there were red colors in the North where they expected to see blue and blue in the rural South where they expected to see red. Their first answer was that red in the North meant that instead of the population increasing at a greater rate from 1930 to 1960, it meant that it was increasing at a slower rate. However, when she asked them how they knew it was increasing at all, they realized that they were assuming that the population had increased in both periods, but they didn't actually know. Ms. Rivera suggested that they check that assumption when they returned the next day. After saving their new data and taking notes about what they had done, Gary and Latonia left for the day.

Investigation Day 2

When they returned the next day, Gary and Latonia decided to look at the other two data values they had calculated, "1900–1930 growth" and "1930–1960 growth" (Figure 4.2). Once again, they were surprised to see a patchwork quilt instead of the consistent patterns they had expected. Without Ms. Rivera's prompting this time, they asked themselves what they had expected to see and whether they saw it anywhere. For both images, they had expected to see a broad pattern of positive (blue) values in the North and urban areas, and a pattern of negative (red) values in the rural South. Since they expected the population trends to accelerate in the later period, they expected to see darker shades of red and blue in the 1930–1960 data and lighter shades in the 1900–1930 data. As they had the day before, they saw some evidence of the patterns that they expected in the Upper Midwest, but not in the South.

When they examined their assumption from the day before, they discovered that there were counties where the African American population didn't grow in both periods and others where it didn't decrease in both periods. When they tried to figure out what the "Change in growth" they had calculated the day before meant for a county whose population grew in the first period and shrank in the second, they became confused. They

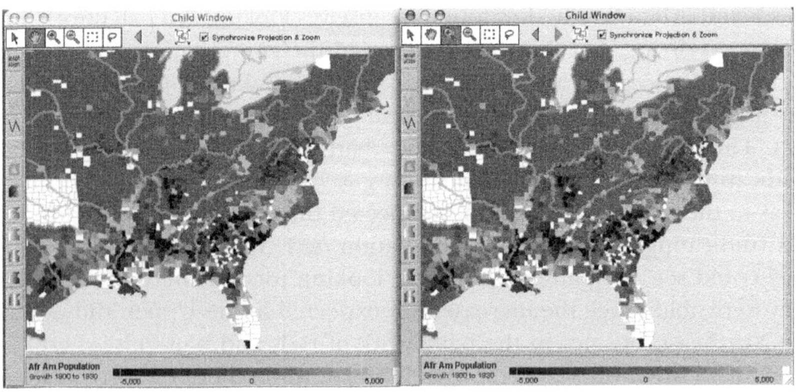

Figure 4.2. A comparison between the growth in African American Population 1900 to 1930 (left) and 1930 to 1960 (right). Increasing population in lighter shades (blue); decreasing in darker shades (red).

understood that a county with growth followed by a loss would have been colored red in the previous day's map, but they didn't think they could interpret that red as meaning the same thing as the red of a county whose population decreased in both periods, but decreased more in the later period. From across the room Ms. Rivera could tell that Gary and Latonia were slipping from confused to discouraged, so she came over to see what was going on.

When they explained that they didn't think they could treat the two different causes of red the same way, Ms. Rivera gave them a suggestion. She suggested that they divide the counties into different categories, so that they could construct maps showing just the counties of a particular type. They didn't understand what she meant at first, so she advised them to create a map where red only had one interpretation.

With that extra hint, Gary and Latonia decided to create a map showing just the counties whose African American populations decreased in both periods. They knew that red in that map would mean that the county had experienced an acceleration in population loss. If their original hypothesis were true, then most of the counties in the rural south would appear as red on their map. If counties in the rural South didn't appear as red on that map, that would mean that those counties did not agree with their hypothesis. They created the map by selecting just the counties whose growth was negative in both time periods (Figure 4.3). When they looked at the map, they were chagrined to see that most of the South did not fall into the cate-

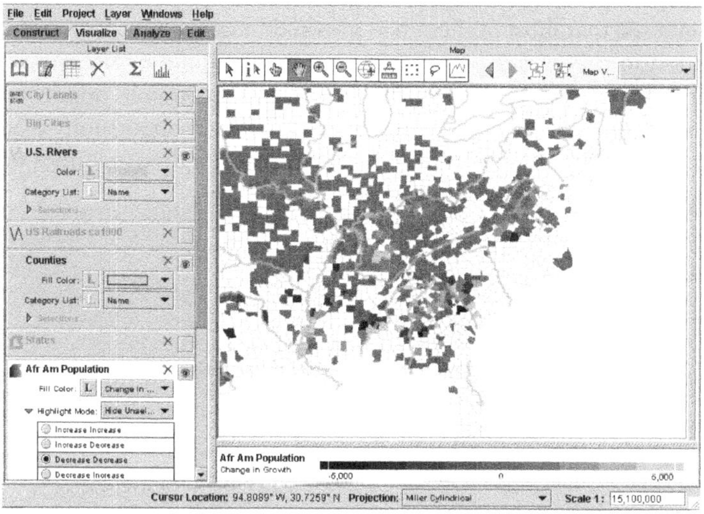

Figure 4.3. Change in growth between 1900–1930 and 1930–1960 for counties whose African American Population decreased in both periods. Other counties appear in white.

gory of counties that decreased in both periods, calling a major component of their hypothesis into question. They put their hands up immediately, and Ms. Rivera made her way over to see what they had found.

While Ms. Rivera agreed that this map cast serious doubt on their hypothesis, she reminded them that they had seen the pattern they expected in some places on previous maps and encouraged them to continue to break down their data into categories and see what they could find in the other categories. She also encouraged them to add cities to their map, so they could distinguish between urban and rural counties. They agreed to follow her suggestions the next day.

Investigation Day 3

To complete the process of categorizing data they had started the day before, they constructed a table similar to Table 4.1 and then queried the data to make a separate map for each category.

When they looked at the "Increase–Increase" subset (Figure 4.4), they were once again able to see the pattern their hypothesis predicted in the northern states. Most counties in the North belonged to the Increase–Increase subset and most of them showed the pattern of accelerating growth (blue) their hypothesis predicted. In the South, they saw both accelerating and decelerating growth. By adding cities to their map (Figure 4.4), they were able to see that most of the counties in the South with accelerating growth (blue) contained cities, as predicted by their hypothesis. When they looked at the individual counties with decelerating growth in the South, which were primarily rural, they could see that most of them had grown dramatically in African American population in the earlier period and experienced little or no growth in the later period. The students speculated that some economic opportunity had drawn them to those areas in the early years, but that opportunity reached its limit, so the growth stopped after 1930. They decided these counties would be interesting to study in the second phase of the project.

TABLE 4.1
Categories of Counties

Name	1900–1930 AfAmPop	1930–1960 AfAmPop
Increase–Increase	Increasing	Increasing
Increase–Decrease	Increasing	Decreasing
Decrease–Increase	Decreasing	Increasing
Decrease–Decrease	Decreasing	Decreasing

Note: Four categories of counties representing the different possible combinations of increasing or decreasing African American population in the time periods 1900–1930 and 1930–1960

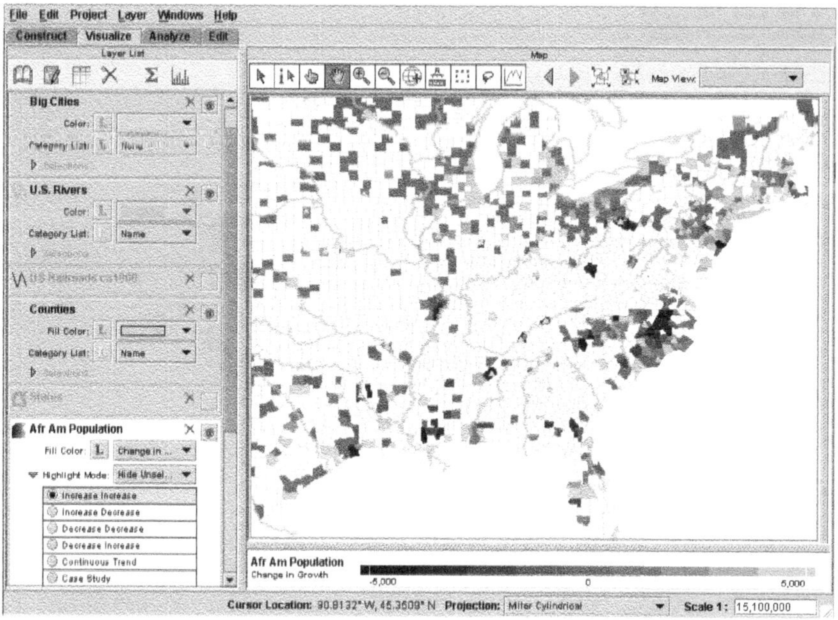

Figure 4.4. Change in growth between 1900–1930 and 1930–1960 for counties whose African American Population increased in both periods. Triangles are cities.

At this point, Gary and Latonia turned to the subsets of the data that were not accounted for in their hypothesis at all: the counties that reversed their trend of growth or loss after 1930. Beginning with the counties that increased and then decreased (Figure 4.5), they observed that these counties were mostly in the South, with a noticeable cluster in the North, in a region that Ms. Rivera identified for them as the Appalachians. With cities added to the map, they were able to see that nearly all of these counties, both North and South, were rural. They concluded that these counties were essentially consistent with their hypothesis as they were rural counties whose population trend changed in the direction their hypothesis would predict. However, as with the counties that increased then stabilized that they'd looked at earlier, they did not have an explanation for why the number of African Americans in rural counties might have increased in the earlier period, particularly in the South. They decided to add this group of counties to the set they wanted to study in the second phase of their project.

When they moved on to the counties whose African American population decreased then increased (Figure 4.6), they noticed a pattern of many counties in the North and few in the South belonging to this category. They also noticed that most of the northern counties were white or light blue indicating a small change in growth, which they concluded meant they had neither gained nor lost much population in either period. In the South, where there

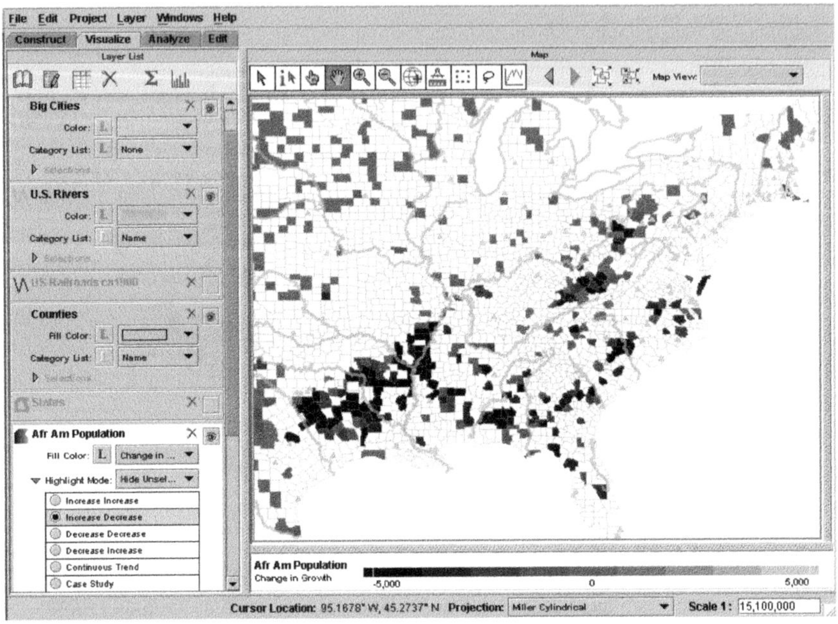

Figure 4.5. Change in growth for cities that increased in African American Population from 1900–1930 then decreased from 1930–1960. Triangles are cities.

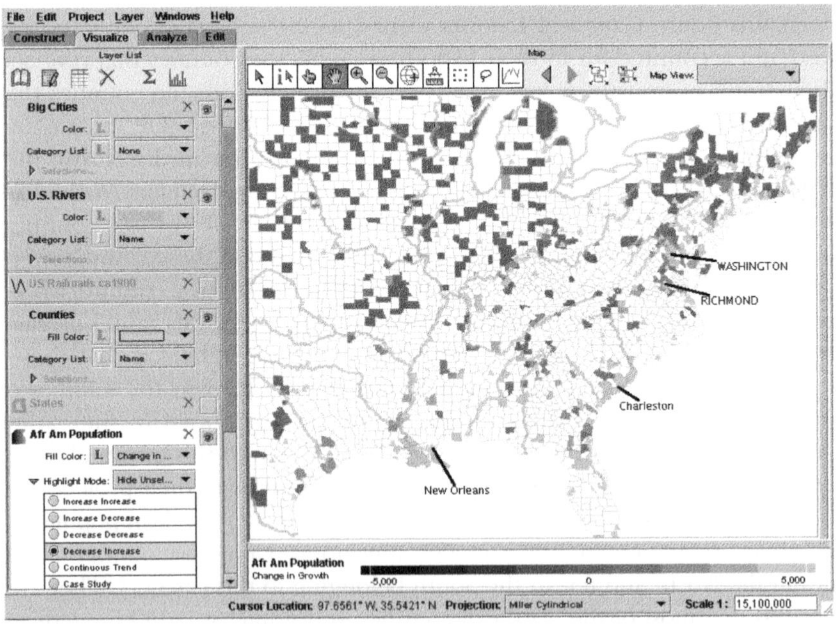

Figure 4.6. Change in growth for cities that decreased in African American Population from 1900–1930 then decreased from 1930–1960. Triangles are cities.

were fewer counties in this category, they noticed that most of them were darker shades of blue, indicating a change of several thousand individuals. When they added cities to their map, they were able to see that several clusters of darker blue counties were in the immediate vicinity of cities. When they identified these cities, they were New Orleans, Charleston, Richmond, and Washington, DC. They speculated that these areas really were following the trend of migration to cities in the South, and that the result of that trend was the transformation of those counties from rural to urban (or suburban). That meant that they would have lost African American population to the North or to the cities in the earlier period when they were primarily rural, but then they would have gained population in the later period as they became urbanized. Gary and Latonia decided to add these specific areas to their research in the second phase of their project as well.

Investigation Days 4 and 5

At the beginning of Day 4, Gary and Latonia met with Ms. Rivera to explain to her what they had found. She told them that they were ready to write up their results and move on to the second phase. She encouraged them to spend the remaining two days on this phase of the project repeating all of the analyses they had conducted to make sure that their results were correct and to record their process for their final presentation. During this time, they constructed a document with a step-by-step description of what they had done and what they had found, illustrated with the maps and graphs they had constructed. At the end of this document, they included the following summary of their findings:

1. Our prediction was true for most of the counties in the Upper Midwest and Northeast and some counties in the South that contained cities. Their populations of African Americans all grew faster between 1930 and 1960 as predicted. Some of the counties in the South lost population more quickly as time passed, as we predicted. But many did not.
2. Our prediction was not true in many counties. Some had slowing growth, slowing loss, or a change in direction. Also, many counties didn't grow or shrink very much at all.
3. In the South, there were rural counties that grew and then stayed the same or lost. We think something must have attracted people there that then slowed down or went away.
4. In the South, there were places where counties around cities experienced loss at first then grew, for example: New Orleans and Charleston. We think this was because those counties became part of the urban area as the cities grew.

We are interested in studying the places in (3) and (4) to see if we can find the causes for why they go against our hypothesis.

Key Features of the Investigation

We now move to a discussion of the role that GIS played in this history investigation and how the features of My World support investigations of this type by novice GIS users. Several aspects of this investigation are significant for this discussion.

The Investigation Involved Multiple Steps, Including the Rejection of Hypotheses and Formulation of New Ones

While there was only one hypothesis under investigation, it required a number of steps to complete, most of which could not have been anticipated or planned in advance. Helping students develop the ability to conduct and monitor such investigations is a widely held goal for social studies (Geography Education Standards Project, 1994; National Council for the Social Studies, 1994; National Research Council (NRC), 2006). In this investigation, the data did not fully support the students' hypothesis. Falsifying a hypothesis requires students to attend to all evidence and systematically evaluate their hypotheses in light of that evidence. Developing these skills is clearly a challenge, as evidenced by research on metacognition, self-regulation, and evaluation of evidence (Allen, Startkiewitz, & Donovan, 1983; Carey, Evans, Honda, & Unger, 1989; Kuhn, Amsel, & O'Loughlin, 1988; Schauble, Glaser, Duschl, Schulze, & John, 1995; Shute, Glaser, & Raghavan, 1989). Nevertheless, recent research conducted at Northwestern has shown that software tools can support students in developing these skills (Kyza, 2004; Loh et al., 2001).

The Investigation Involved Computational Analysis

Throughout the investigation, the students alternated between using operations to calculate mathematical relationships or query for subsets of data and inspecting maps, tables, and graphs visually. In a powerful computational environment like a GIS, the implementation of each of those operations can be complex and places a substantial cognitive demand on students that can make it difficult for them to keep track of the larger context of their investigation. In classroom trials of different GIS environments, we have observed the tendency for students to get so caught up in the "trees" of implementing actions that they lose track of the "forest" of why they are doing those actions and what to do with the result.

The Investigation Involved Quantitative Analysis of Data

In contrast to the typical K–12 social studies investigation, this investigation involved the execution of mathematical operations and interpretation of representations of quantitative data. While the primary operation (subtraction) is simple, the interpretation of change in rates adds an order of complexity. The investigation also involved interpretation of histograms, data tables, and simple statistics. The quantitative skills required by this investigation are all represented in the K–8 math curriculum, yet students often struggle to develop them in the decontextualized math curriculum and often are not prepared to apply them in the context of a social studies class. On the other hand, learning to apply quantitative data analysis skills is an important objective for social studies instruction.

The Investigation Involved Database Queries

To make sense of the patterns in the data, it was necessary for the students to segment the data and recombine it in various ways using database queries and Boolean set operations. These skills are not typically taught as part of the formal curriculum in schools, although they are critical for inquiry across the social and physical sciences.

The Investigation Integrated Computational Analysis with Visual Interpretation

In order to make sense of the data and their analyses of the data, the students engaged in visual inspection of visual representations that included maps, graphs, and tables. Further, they modified the representation in numerous ways to make significant patterns more apparent and reduce visual distraction. While the ability to generate spatially oriented visualizations is one of the recognized benefits of GIS, the interpretation of these representations can be challenging, and the manipulation of aspects of the representation (e.g., symbolization) to expose patterns is a critical skill (National Research Council, 2006).

Design of GIS Tools to Support Investigations

The preceding paragraphs show how the range of features provided by a GIS environment can support a meaningfully complex investigation like the one described in the vignette above. They also show the challenges inherent in supporting such an investigation. The challenges are best characterized as an interaction between student ability and tool complexity.

Students

Students, as with any tool users, come to their use of the tools with both a fixed set of current abilities and the capacity to extend their abilities by learning. While their limited abilities represent a challenge for design, their ability to learn is an opportunity. Our goal in engaging students in complex, authentic tasks is to develop and extend their abilities. Therefore, our design efforts are intended to allow students to use their existing abilities to enter into the investigations, and to use the investigations as opportunities to extend their abilities.

Tools

GIS, as with any type of tool, presents challenges to users on two levels. One is that certain operations or tasks may be either conceptually or logistically challenging. A conceptually challenging operation might be a statistical analysis that requires a conceptual understanding of statistics that the student has not achieved. A logistically challenging operation might be a query that requires students to specify many different parameters. The second is that the achievement of a particular goal with a tool may be challenging because of its complexity. That is, it might require a number of steps, each of which is manageable on its own, but present challenging complexity in combination. The goal of the My World design has been to address these challenges in order to support investigations like the history project described above.

Supportive Features in My World GIS

The design of My World is driven by the notion that the features of GIS should be accessible to students as young as middle school age. This goal requires features that reduce both the conceptual and logistical challenges of GIS operations and provide support to manage the complexity of extended investigation activities. In this section, we highlight a few of the features in My World that have been designed to address these challenges and show the role they could play in the investigation described above.[1] We divide these into two categories: 1) Structured, supportive interfaces to operations, and 2) Persistent, named products of investigations

Structured, Supportive Interfaces to Operations

A hallmark of the ongoing My World development process is the strategic selection of functionality combined with the careful design of interfaces for that functionality. In the strategic selection process, the design team evaluates the general usefulness and power of each possible function. The dual goals are to keep the overall range of functions as small as possible to reduce interface complexity while making sure that every function included is as accessible as possible. One of the best examples of this approach is the design of the interface for analysis operations. The analysis operations are accessed through a special interface that divides the available operations into categories that are meaningful to novice users (Figure 4.7, left).

Once users select a specific operation, they are presented with a form to fill out to specify the details of the operation (Figure 4.7, right). This form

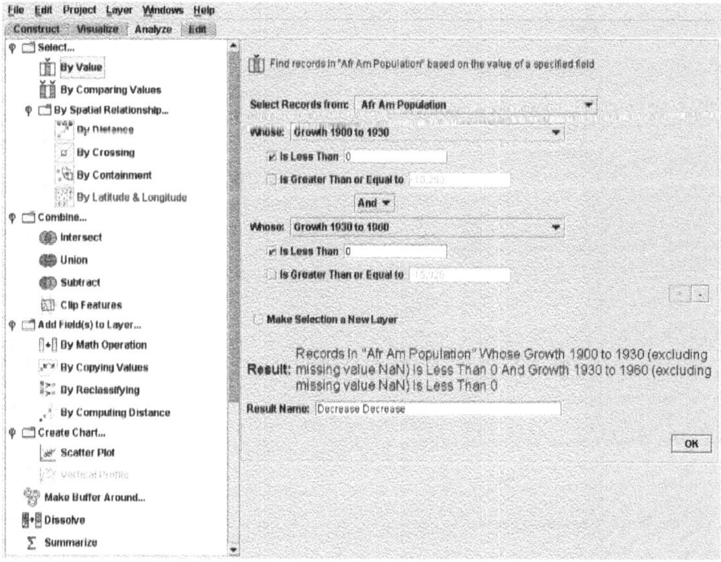

Figure 4.7. The My World interface for conducting data analysis, showing the menu of options at the right, and the form for selecting a subset of records based on their values ("Select by value") at right.

provides as much structure as possible, careful wording to minimize the use of technical terms, and plain text descriptions of the operation (Figure 4.7 upper right) and the result (Figure 4.7, lower right). Finally, to avoid errors caused by mistyping, this form maximizes the use of menus to make selections and minimizes the use of typing. The sentence describing the result (Figure 4.7, lower right) serves as a place to confirm that the correct selections have been made before executing the operation. All selections stay in place after the operation is executed so that the user can return later to make sure she entered the right values.

By way of comparison, a user who wants to query a database within common professional desktop GIS software would construct that query by typing a combination of algebraic and Boolean notation. This process requires users to master the formalism of a query language, remember the arguments (and order of arguments) necessary to perform an operation, manage syntax involving a balance of parentheses and brackets, and avoid mistyping. This interface creates large conceptual and logistical challenges that often limit students' analyses and require unreasonable amounts of classroom time for learning formalisms and interfaces and conducting analyses.

The examples illustrated throughout the vignette involved the use of the My World GIS analysis interface to calculate differences and conduct database queries. My World GIS makes quantitative and database analyses

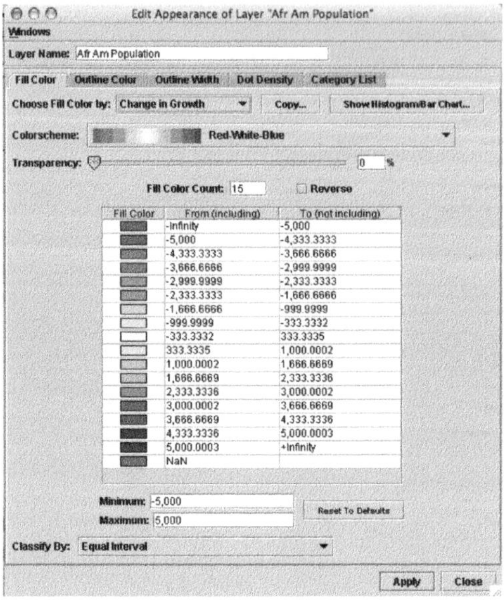

Figure 4.8. Interface for customizing the appearance of a layer in My World using a structured form.

accessible to users who are relative novices. This analysis interface addresses the conceptual challenges of analytical operations by providing plain text descriptions of the operations, accompanied in some cases by graphics, and by breaking the operation down into individual, comprehensible steps. It minimizes logistical challenges through the reduction of interface complexity by organizing choices into categories, using everyday language wherever possible, and structuring the interface to prompt users for all the required information and to minimize errors from inattention. While in some cases the design of these interfaces was based on the prior experience of the development team, many of the design decisions have been informed by user studies in lab or classroom settings. My World has been designed to provide similar interfaces to structure and support the customization of visual representations thus responding to the third key feature of the investigation listed earlier, *integration of visual interpretation.*

Persistent, Named Products

To address the challenges associated with complex, multi-step investigations, My World provides extensive support for saving, naming, and returning to intermediate work products. Prior research at Northwestern has shown the value of retaining and indexing intermediate work products (Kyza, 2004; Loh et al., 2001) such as 'Mapviews' that are generated as students build their maps to support student investigations. This is particularly important for learners who lack well-developed metacognitive abilities and/or are working under a substantial cognitive load because they are facing multiple conceptual and logistical challenges simultaneously. In some cases, My World retains and names intermediate products automatically, thus further reducing the burden for metacognition. In other cases, My World offers users the option to save an intermediate product through a "save" button displayed in a prominent location.[2]

The most prominent example of persistent, named products in My World is what we call *selections*. A selection is a subset of the records in a data set (or rows in a data table) that have been identified by the user through a query or direct interaction with a map. Whenever a user creates a selection using My World it becomes a persistent, named object. If the user creates a selection through a database query, she has the option to name it herself (e.g., "Increased 1900 to 1930") or to allow the software to create a default name based on the operation (e.g., "AFRPOP1930 >= AFRPOP1900") (Figure 4.9)

The names of the existing selections for a particular layer are displayed under the name of the complete data set in the layer list, and the user can choose which selection to highlight (if any) and can perform any operation on a selection that can be performed on a the entire data set (See Figure 4.10).

Selections also support their own operations, such as set combinations (union, intersection, and difference), and can be deleted if desired. From a logistical standpoint, the presence of persistent, named selections makes it substantially easier for novices to create complex queries. Rather than needing to construct a complex query all at once (e.g., the union of counties whose AfAmPop increased and then decreased and counties whose AfAmPop decreased then increased), it enables users to construct the

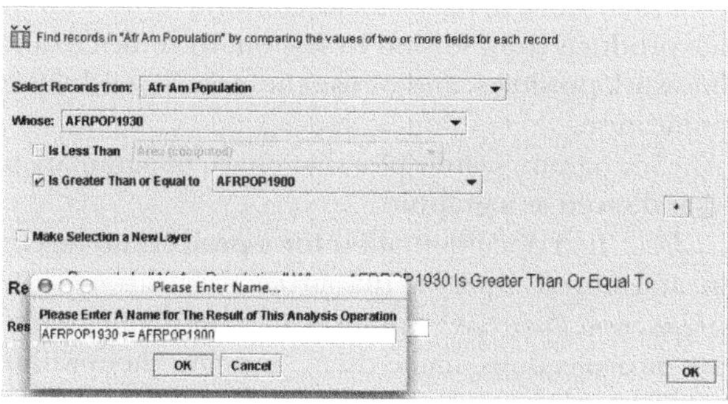

Figure 4.9. The interface for naming a "selection"—a persistent subset of a data set—created through a database query.

Figure 4.10. Layer interface (left) showing a list of existing selections (subsets) for that data layer. Currently the selection entitled "Reversing Trend" is highlighted using the "Hide Unselected" option.

query in individual steps (e.g., first counties who increased then decreased, second counties who decreased then increased, third the union of the first two). From a conceptual standpoint, the representation of a selection as a "layer within a layer" can help learners to understand what constitutes a selection. From the perspective of managing investigations, persistent, named selections allow learners to return to prior queries to review them or take a new action on them. Furthermore, students have a visible artifact to which they can refer in communications about an investigation with co-investigators or with outsiders.

Other examples of persistent, named entities in My World are:

- *Mapviews.* A mapview is a specific customized view of data that is specified in terms of extent and scale of the map, the layers that are visible, and the colorscheme or other symbolization for those layers.
- *Projects.* All of the persistent, named items created in a session can be saved in a single project file along with all of the data in a mapview, so that the state of a student's work can be maintained from session to session. This allows students to build and maintain an ongoing record of their work in order to: a) continue work with no need to recreate products from session to session, b) return to and reflect on previous work products, and c) use these work products to support communication.
- *Plots.* Any histogram, scatter plot, bar chart, or summary table can be named and saved as a graphic.
- *Text windows.* Text documentation for a project can be created, named, and stored as part of that project.
- *Colorschemes and Projections.* Sophisticated users can create new colorschemes or projections, name them, and save them within a project.
- *External files.* Links to external files can be created, named, and saved. These can be linked to specific records in a layer, or they can be associated with the entire project.

The ability to name and save work products within and between sessions is a critical feature for users who are still developing the skills for planning, conducting, and monitoring investigations. It reduces the memory burden on them, and it accommodates the nature of school schedules where most work is conducted in a series of brief, interrupted sessions, rather than in the extended sessions in which most professional users of GIS work.

FINAL THOUGHTS

The underlying message of this chapter is that GIS software designed with an understanding of students, the learning process, and the educational context can help bring the promise of geospatial data analysis to social studies classrooms. While My World is a relatively new software environment, first published in 2004, the hypothesis that it can provide access to the analytical capabilities of GIS has been borne out in feedback to the design team from dozens of middle school through college teachers. While actual usage figures are hard to obtain, hundreds of copies of the free trial version are downloaded from the My World website each month, and the publisher has sold licenses numbering in the tens of thousands.

During the 2005–2006 school year, we conducted an initial comparative study of My World with a widely-used commercial GIS application.[3] Preliminary data analysis suggests that students using My World spend more time discussing the content of their investigations and less time consumed with learning the software than students using professional GIS tools. Similarly, students using the professional GIS tool experienced more logistical problems with the software that required them to spend substantially longer periods of time waiting for assistance or figuring out how to solve the problem than My World users. Although these results are preliminary, they suggest that My World offers advantages over professional GIS software in educational settings. Further analysis is needed and is underway to further investigate the use of My World in secondary school settings.

Software that addresses logistical and conceptual challenges is only a piece of the puzzle of engaging students in meaningful investigations with geospatial data. There remain many questions about teacher preparation and curriculum design that need to be addressed in order to realize the promise of GIS in the social studies classroom. One particularly thorny issue is raised by the vignette used to illustrate this chapter. The vignette depicted an open-ended and student-directed investigation involving students who possessed appropriate prior experience with GIS and data-driven inquiry and a teacher skilled in guiding inquiry. We developed this example because it represents the possibilities for students' use of inquiry techniques with GIS in school and in their lives. Although it would not be unusual to see GIS or other inquiry applications being used to address a question with a similar level of complexity in today's classrooms, it would be very unusual to see a student-directed investigation of a hypothesis generated by the students not the instructor, as we described in the vignette. The typical implementation of inquiry in American classrooms consists of following a set of step-by-step instructions with prompts to interpret, reflect, and communicate results. As GIS and similar inquiry tools become more accessible through advances in technology and design for learners, educators have the opportunity to help students move toward open-ended inquiry, rather than being satisfied with teacher-directed activities. Only when students have learned to plan, execute, monitor, and re-plan their own investigations can we say we have taught them to engage in meaningful inquiry.

NOTES

1. Throughout this section, there is an implicit comparison between My World and GIS tools designed for professional users. Most of My World's functionality can be found in any of the most widely used desktop GIS tools, though the interfaces are often dramatically different. For reasons of brevity, we have not provided illustrations of those tools. Readers who do not have prior experience with those tools can find screenshots in most curriculum guides for classroom GIS use or books on GIS use in the classroom.
2. In prior research, we have also observed the benefits of providing users with places to annotate their work and prompts for reflection (Kyza, 2004; Loh et al., 2001), which we have not yet been able to implement within My World.
3. In order to avoid a focus on the strengths or weaknesses of any individual tool, we have a policy of not naming the specific tools used in our comparative studies. Based on our analysis of the existing tools, we expect we would obtain similar results for any professional GIS tool.

ACKNOWLEDGEMENTS

The authors gratefully acknowledge the contributions of the many teachers, students, and members of the GIS in education community who have shared their ideas and experiences with us in the development of My World. We also thank Josh Radinsky for sharing his GIS for History data with us. This material is based upon work supported by the National Science Foundation under Grants No. ESI-0352478, DUE-0333529, EIA-0085946, and ESI-9720687. Any opinions, findings, and conclusions or recommendations expressed in this material are those of the authors and do not necessarily reflect the views of the National Science Foundation.

My World GIS and *My World* are trademarks of Northwestern University. My World GIS is distributed by PASCO Scientific (http://www.pasco.com/myworld). My World was programmed by Eric Russell.

REFERENCES

Allen, R. D., Statkiewitz, W. R., & Donovan, M. (1983). Student perceptions of evidence and interpretations. In J. Novak (Ed.), *Proceedings of the international seminar: Misconceptions in science and mathematics* (pp. 79–83). Ithaca, NY: Cornell University.

Carey, S., Evans, R. M., Honda, E. J., & Unger, C. (1989). An experiment is when you try it and see if it works: A study of grade 7 students' understanding of the construction of scientific knowledge. *International Journal of Science Education, 11*(5), 514–529.

Edelson, D. C., Gordin, D. N., & Pea, R. D. (1999). Addressing the challenges of inquiry-based learning through technology and curriculum design. *Journal of the Learning Sciences, 8*(3&4), 391–450.

Geography Education Standards Project. (1994). *Geography for life: National geography standards 1994.* Washington, DC: National Geographic.

Kuhn, D., Amsel, E., & O'Loughlin, M. (1988). *The development of scientific thinking skills.* San Diego: Academic Press.

Kyza, E. (2004). *Understanding reflection-in-action: An investigation into middle-school students reflective inquiry practices in science and the role that software scaffolding can play.* Unpublished Ph.D. Dissertation, Northwestern University, Evanston, IL.

Loh, B., Reiser, B. J., Radinsky, J., Edelson, D. C., Gomez, L. M., & Marshall, S. (2001). Developing reflective inquiry practices: A case study of software, the teacher, and students. In K. Crowley, C. Schunn, & T. Okada (Eds.), *Designing for science: Implications from everyday, classroom, and professional settings* (pp. 279–323). Mahwah, NJ: Erlbaum.

National Council for the Social Studies. (1994). *Expectations of excellence: Curriculum standards for social studies.* Washington, DC: National Council for the Social Studies.

National Research Council (NRC). (2006). *Learning to think spatially: GIS as a support system in the K–12 curriculum.* Washington, DC: National Academy Press.

Radinsky, J. (this volume). GIS for history: A GIS learning environment to teach historical reasoning. In A. J. Milson & M. Alibrandi (Eds.), *Digital geography: Geospatial technologies in the social studies classroom.* Charlotte, NC: Information Age.

Schauble, L., Glaser, R., Duschl, R. A., Schulze, S., & John, J. (1995). Students' understanding of the objectives and procedures of experimentation in the science classroom. *Journal of the Learning Sciences, 4*(2), 131–166.

Shute, V. J., Glaser, R., & Raghavan, K. (1989). Inference and discovery in an exploratory laboratory. In P. L. Ackerman, R. J. Sternberg, & R. Glaser (Eds.),

Learning and individual differences (pp. 279–326). New York: W. H. Freeman and Company.

TERC. (1995). *First national conference on the educational applications of geographic information systems (edgis) report.* Cambridge, MA: TERC.

CHAPTER 5

GIS FOR HISTORY

A GIS Learning Environment to Teach Historical Reasoning

Josh Radinsky
University of Illinois at Chicago

GIS brings to the history classroom an exciting new palette of possibilities and challenges. For teachers who are looking for ways to engage their students in *doing* history (Levstik & Barton, 1996)—i.e., investigating, examining, and interpreting historical information sources—GIS can make authentic investigation of the past possible in unprecedented ways. With GIS maps, students can visualize the complexity of the American demographic, political, and economic landscape, making subtleties of data patterns visible as colors and shapes. Change over time can become an object of scrutiny, accessible for observation and discussion through a visual language. Students can explore geographic relationships among people, places, events, and human-made and natural resources, actively inspecting them from multiple perspectives.

However, for all the advantages GIS offers to educators, there are also significant challenges it brings for historical reasoning. The uncertain and incomplete nature of the historical record is easily misunderstood in the clean interface of a computer map. The wealth of information in a GIS database may give students a false impression that "it's all there," or that

Digital Geography: Geospatial Technologies in the Social Studies Classroom, pages 99–117
Copyright © 2008 by Information Age Publishing
All rights of reproduction in any form reserved.

they are working with the "raw facts." Complex visual data can be overwhelming and difficult to interpret. Correlations among data points or objects in a map can easily be misinterpreted as evidence of meaningful relationships, where none may really exist.

Students need to learn to make sense of complex visual data displays, "reading" GIS maps and making appropriate inferences and observations. They need to develop skills and habits of historical inquiry, generating questions, seeking evidence, and building an argument. They need to learn to reason appropriately with data, avoiding fallacies in reasoning that are common among youth and adults alike, and connecting abstract representations of data back to their referents: real people living in remote times and places.

These learning objectives underlie the design of *GIS for History* (http://www.gisforhistory.org), a public-use learning environment funded by the National Endowment for the Humanities as a tool for teaching American History in grades 6–16. The website includes interactive GIS maps for studying historical census data (1790–2000); lesson plans with suggestions for teaching specific historical themes with the GIS; and links to selected online historical documents that are chosen to help students ask and answer historical questions relevant to the census data in the maps. This chapter examines the design of *GIS for History*, using examples from pilot studies to illustrate how these kinds of historical reasoning play out in high school history classrooms.

Historians and GIS

GIS has become a widely-used analytical tool in many disciplines, including geophysics, anthropology, demography, and many others. While the place of GIS in professional historical inquiry is still one of uncertainty, a small group of innovators has begun to identify a fascinating range of methods and arguments that bring new historical insights (Knowles, 2002). For example, Geoff Cunfer's studies of the Dustbowl (2002; 2005) have challenged a range of widely held assumptions about the causes of the Depression-era plains dust storms. Cunfer's analysis uses GIS maps representing population, crop, soil, and precipitation data with analyses of a range of historical documents from the period. His findings challenge the accepted historical narrative that over-farming of croplands led to the devastating dust storms, showing how broader scales of geography and time, analyzed using a GIS, reveal that the dust storms of the 1930's were more likely attributable to weather patterns than to human causes.

Similarly, the Knowles, Farrell, and Roush (2004) study of the Civil War battle at Gettysburg used GIS to construct a line-of-sight "viewshed" analysis of the battlefield, from which she was able to infer what information would have been available to commanders as they prepared for and engaged in battle. Her analysis sheds new light on military decisions by Confederate and Union commanders at key moments during this critical engagement of the Civil War, suggesting not only logistical factors but also subjective, perceptual experiences of commanders that may well have played a significant role in the unfolding of events.

Other historians have used GIS for studies of topics ranging from 12th century Chinese administrative systems (Bol, 2004) to comparisons between plank roads and railroads in 19th century Pennsylvania (Healy, 2006). These pioneering works of historical research have begun to establish a wider acceptance of GIS as a tool for historical inquiry (Young, 2006). It appears that GIS as a tool for historical research, while in its infancy, is beginning to emerge as a valued methodology. As tools improve and shared databases develop, the world of historical GIS is likely to grow in the coming years, as it has in so many other domains.

GIS AS A TOOL FOR TEACHING HISTORY

Although its classroom use to date has been most common in environmental science, GIS technology has great promise as a tool for teaching history. GIS has the potential to help history teachers provide their students with rich and complex sources of information, such as historical census data, for inquiry projects. With appropriate guidance from an instructor, such data can give students opportunities to construct deeper understandings of complex changes over time, including topics such as industrialization, urbanization, suburbanization, migration, immigration, and numerous other core themes of American history. Logistical challenges for instructors include being able to give their students access to appropriate data for a particular instructional theme, and being able to monitor and support students in making sense of such complex data.

THE GIS FOR HISTORY WEBSITE AND CURRICULUM

GIS for History was created to put into students' hands a number of newly-created digital resources that are combined in a format that can be easily used for historical investigations. The site's interactive maps use historical county and state boundary files for every census year since 1790, recently created by the National Historical GIS project (http://www.nhgis.org). These shape files are linked to selected census data variables from each year, acquired from digital archives such as the Integrated Public Use Microdata Series (http://usa.ipums.org/usa/). In addition to these geographic and numerical sources, *GIS for History* also provides links to selected documents from a number of excellent online collections of primary source documents that have emerged in recent years, including History Matters (http://www.history matters.gmu.edu/), American Memory (http://memory.loc.gov/ammem/), Gilder-Lehrman (http://www.gilderlehrman.org/), and others.

The goal of *GIS for History* is to enable students to investigate key topics in American history using a combination of census data maps and historical documents. The census data GIS maps provide rich contextual information about the historical period, and a means of visualizing broad changes over time and space. The primary-source documents provide multiple perspectives on historical events, and evidence of the agency of historical actors. Together, the GIS and the documents can be used by students to generate inferences about historical processes and changes in American society, and then to look for evidence to support or counter their historical arguments.

Design of the Web Site

The balance between documents and census data is meant to help students generate and investigate authentic questions about American history. This balance is embodied in the design of the *GIS for History* website. Once students access the site and select a project, they are provided with three tabs for site navigation: OVERVIEW, MAP, and DOCUMENTS. At any time students can move from one mode of work to the other, making it easier to make connections among multiple sources of information.

GIS for History was created as a learning environment for classroom instruction, not as a general-use GIS map for viewing historical census data. For this reason, the first choice one must make upon visiting the site is to select one of the curriculum units (see Table 5.1). The site is organized around lesson plans, written by curriculum designers and high school history teachers, addressing specific topics in American history. They are written at the high school level, but are intended to be adaptable for simpler middle-school assignments or more involved undergraduate projects. Teachers may use individual lessons for a one-class activity, or may use the full materials for extended investigations in which students conduct in-depth research.

The GIS Maps

Each set of lessons has a customized GIS map, with particular years, census data variables, and additional map layers chosen for the topic of that

TABLE 5.1
Curriculum units provided as part of *GIS for History*

Curriculum unit	Years	GIS map layers	Document collections
U.S. Expansion	1790–1970	• Total population by state, county • Largest cities by rank and population	• Letters from presidents Jefferson, Jackson • Documents evidencing Anglo migrants' conceptions of the West
The First Census: America in 1790	1790	• Total population by state • Race (White, African American) by state • Enslaved and slaveholding status by state • Nationality (8 categories) by state • Largest cities, by rank and population	• Debates within the Constitutional Convention • Accounts and text of the Great Compromise, 3/5ths Compromise • Image: signing of the Constitution
Slavery in America	1790–1870	• Total population by county • Race: African American, White, enslaved, free • Change in enslaved population, county by decade • Largest cities by rank and population • Major railroad lines	• Narratives of enslaved people • Legislation of slavery • Documents for 8 case studies: agency of enslaved people
The Great Migration	1880–1960	• Total population by county • Race: White, African American • Afr Amer population change, county by decade • Industrial jobs and wages by county • Value of crops by county • Largest cities by rank and population • Major railroad lines (to 1900)	• Letters from migrants • Photo archives • Accounts of sharecropping • White racism, North & South • Poetry and other literary links • The Great War
Native American Lands: the Black Hills (map under construction)	1850–1910	• Total population by county • US-Lakota treaty area boundaries • US-Lakota battle sites • US military outposts • Migration trails used by White settlers • Major railroad lines (to 1900)	• The Black Hills: Lakota perspectives • Treaties between US and Lakota, Nakota • Red Cloud's War • The transcontinental railroad • The Dawes Act

curriculum unit (see Table 5.1 for details of each unit). A timeline, with buttons at 10-year intervals, controls the data displayed on the map. When a year is selected on the timeline, the county and state boundaries change to reflect that year's census geographies, as do the other map layers, such as "Biggest Cities" and "Major Railroad Lines." The census data displayed on the map are also controlled by the timeline, so color patterns shift as students select different years. The color schemes for representing historical census data on the GIS maps are adapted from *ColorBrewer* [http://ColorBrewer.org], a widely recognized standard.

Layers of the GIS map can be turned on or off, to show or hide information such as the locations of rivers or cities. Students can zoom in or out to magnify or shrink the map, and they can move the map in any direction by clicking and dragging the mouse. This allows easy access to view any part of the country. A click on any place in the map provides a report of the data available for that location (e.g., census data for the county, state, or city selected). The layer that is highlighted in the legend will provide the available data for the location where the map was clicked. For example, in Figure 5.1 the student has highlighted the "Biggest Cities Population" layer, and the timeline is set to 1880, so a click on Chicago labels the city on the map and shows that the city's population in 1880 was 503,185 people. This number shows up in the map's Query History (bottom right of Figure 5.1), and also displays above the legend for "Biggest Cities Population" to show where in the range of the data that particular city falls.

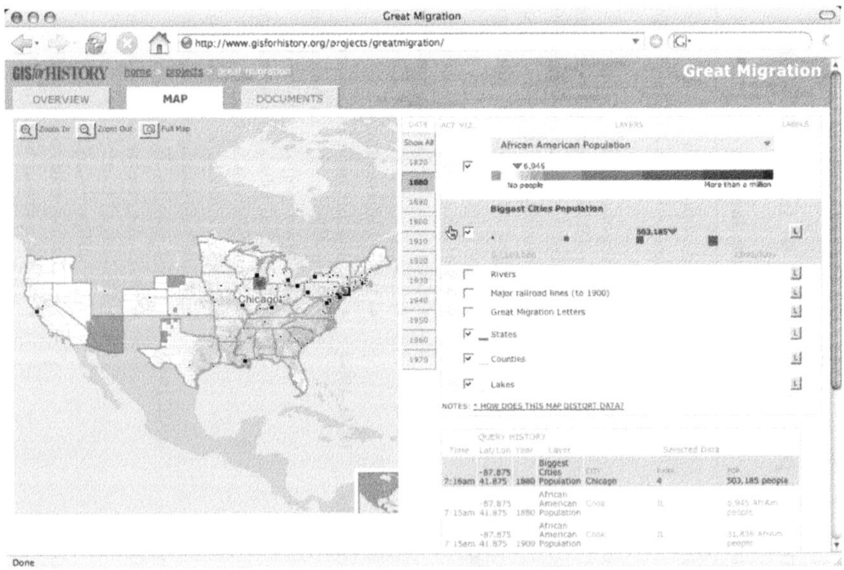

Figure 5.1. The Great Migration GIS map showing African American population by county, and the 100 largest U.S. cities, in 1880.

The information provided after each click is stored in the Query History list, so that students and teachers can easily trace what has been done. The most recent click is at the top of the list, with a trail of all the student's previous actions following below. For example, Figure 5.1 shows a map in which a student has selected the "African American Population" layer for 1900; clicked on Cook County, IL; changed the year to 1880; and then turned on the "Biggest Cities Population" layer. The Query History still displays the data from those previous clicks, showing that the county had 6,945 African Americans living there in 1880, compared with 31,838 twenty years later. Clicking on any previous item in the Query History returns the map back to the state it was in at that time, making it easier for students to keep track of their progress, and for teachers to review students' work and help them make sense of their investigation.

Document Collections

In addition to a customized GIS map, each curriculum unit also includes one or more document collections for students to use. Links to a few selected documents, most of them chosen from archives such as *History Matters* and *American Memory*, are organized for a particular historical theme. For example, in the *Great Migration* unit, documents are provided on themes including sharecropping, the Great War, racial violence, and letters written by African American migrants describing life in the North and South. These linked documents include historical images such as photographs and political cartoons, as well as text documents such as letters, speeches, or legislation. The collections are not meant to be exhaustive libraries on each topic, but rather documents carefully chosen by teachers and curriculum designers to be used in the lessons provided. If teachers wish to include more documents, students do not have to be limited to those chosen on the *GIS for History* web site. Sites like *History Matters* and *American Memory* also provide excellent search tools and document indices, so students can easily conduct further independent research on topics of their choosing.

GIS IN THE HISTORY CLASSROOM: OBSERVATIONS FROM PILOT STUDIES

The ten high school teachers and their students who have used *GIS for History* in pilot studies have shown enthusiasm for the investigations, and have found the GIS tools to be useable and engaging. Some of the whole-class discussions using the GIS maps have been particularly exciting. Several teachers have noted that the students who participate in these GIS discussions often include some who are not the "usual suspects" in the academic discourse of their classrooms: something about the GIS experience seems to bring more students into the process of historical reasoning and debate. For example, during pilot studies one history teacher noted that the first students to speak during class discussions with the GIS were students who rarely played the role of "conversation starters" in his classroom (M. Lauterbach, personal communication, 4/18/2006).

Similarly, our ongoing studies of students using GIS in collaborative small-group work suggest that students who are often quiet with their peers may find avenues for developing increased agency as contributors to inquiry work through the use of visual data (Radinsky et al., 2006). Often these students' initial observations and explanations from the GIS begin in the form of gestures, before emerging as full elaborations of their ideas in speech (Singer, Radinsky, & Goldman, in press). Evaluation studies currently under way will tell us more about specific strengths and weaknesses of the tools, the curriculum, and the instructional approaches used in *GIS for History*. In the meantime, we share here some observations from our classroom experiences to date, and some issues for the teaching of history that have emerged.

Learning to Read Visual Data

Students may talk excitedly about what they see on a GIS map, but they often have difficulty describing it with precision and clarity. Students need to learn to make careful and accurate observations of data patterns. This includes noticing broad patterns of a data distribution, while also seeing subtle variations in those patterns. For example, at the beginning of the Great Migration project the teacher asks students to make observations about a population map: Where did most African Americans live in 1900? (See Figure 5.2, left side.) At first students may just make broad or simple observations: "It's all down there," or "Mostly in the South."

Teachers can help students learn to make such general statements more specific, and also to look more closely at the data. For example, one teacher used an example from a student's work to point out a description of a broad pattern in 1900, and then to point out specific changes to that pattern that show up in 1940 (both shown in Figure 5.2):

> Teacher: [A student said that] most African Americans were living in the South in 1900. That is a good description of what the map shows.

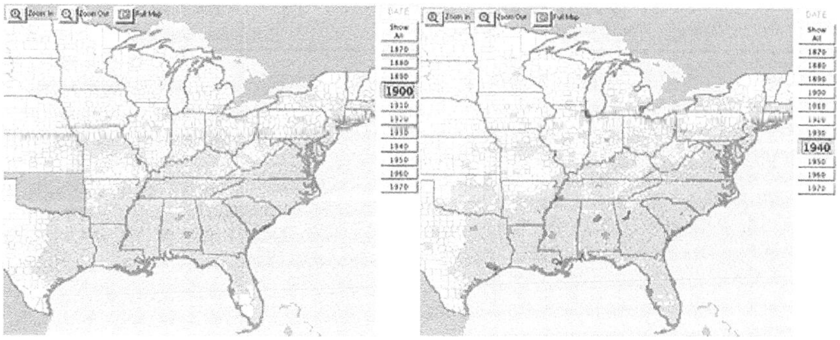

Figure 5.2. African American population by county, 1900 and 1940.

Teacher: ...and then she said, "By 1940 a lot of counties in the North had high African American populations ... like Cook County, Illinois, and Wayne County, Michigan." Very nice examples, and you see these little dark spots that you see around the North. That is what she was talking about.

Student: But there's also dark spots in the South, so that is not necessarily true.

This teacher models both a broad observation of a pattern (most African Americans living in the South in 1900), and then more subtle variations in that pattern in the next map (some individual Northern counties that have high African American populations by 1940). The teacher explicitly connects a visual description of the 1940 data pattern ("little dark spots around the North") to the real-world referents of the data ("Cook County, Illinois, and Wayne County, Michigan"), modeling the kind of specific observations he is asking students to make with the GIS map.

In the above example, a student objects to part of her teacher's description. Her more careful observation of the data pattern ("there's also dark spots in the South") raises an important historical point about the Great Migration—that African American population increased dramatically in certain Southern counties as well as the Northern ones mentioned. Her statement "so that is not necessarily true" is ambiguous, but suggests that she may be reasoning not just about "dark spots" on the map, but also about the meaning of the historical data described by the teacher.

As the class session progresses, this same student goes on to make her observation more specific. As part of a data gathering activity (described below), she copies images from the GIS to show a set of specific observations, and discusses their significance (Figure 5.3). She first shows an example of a Southern in-migration county, Jefferson County, Alabama. She makes her map's Query History show the growth in African American pop-

EVIDENCE: Paste evidence from letters & maps	**ELABORATION:** What does this evidence show?
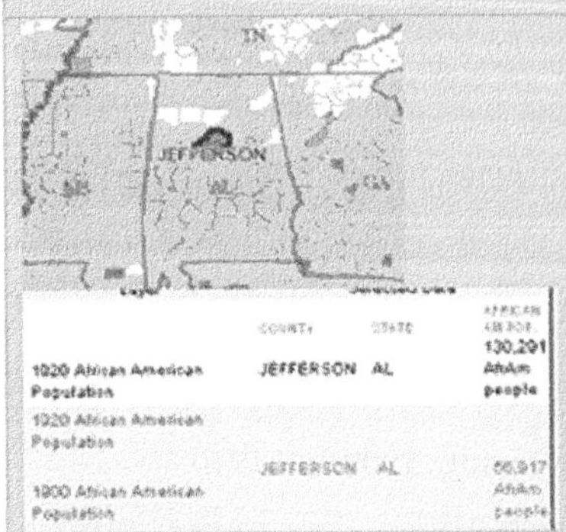	This is a map showing that African Americans were not necessarily moving from the south. Jefferson county has a big city, called Birmingham; it is ranked 99 as one of the biggest cities in 1900 and 36 in 1920.
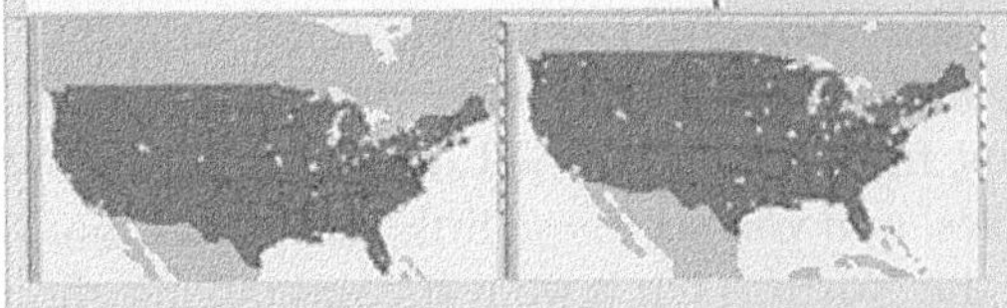	This is a picture showing the big cities in the U.S. in 1900 (left) and in the 1920 (right). This shows that all the big cities grew from 1900 to 1920. Notice that they grow quite a bit from 1900 to 1920.

Figure 5.3. Screen images captured by a student and pasted into a Word document, to describe migration patterns she has observed.

ulation for that county from 1900 to 1920 (increasing from 56,917 to 130, 291, as shown in Figure 5.3). She also points out the presence of a big city, Birmingham, in that county.

She then moves back from this close-up of one county to a wider view, capturing two continent-wide maps to show the growth of big cities nationwide over this same time period (bottom of Figure 5.3). Her observation that big cities "grow quite a bit from 1900 to 1920" is not very specific, but her detailed observations about Birmingham (rising in rank from 99th to 36th largest city in America) and the rise in the African American population of that Southern city's county are more important to the point she is making about intra-South migration.

This example shows the kinds of careful observations students can make with the visual data in a GIS. The challenge for teachers is to help students learn to make such observations, and to describe them clearly and accurately. Most of the students in the pilot classrooms, though not all, learned to make such careful observations of data patterns.

Using Data as Evidence for an Argument

The student in the example described above has begun to move from observation of a pattern (some dark spots in the South) to developing a historical claim (the Great Migration was not entirely South-to-North, but also included migration to Southern cities). Students struggle with this kind of reasoning from data, partly because they must be organized in their work, connecting a series of observations from the data and developing them into a clear narrative of a historical event or process. Teachers using *GIS for History* have supported students in several ways to teach them this process.

One way to support historical reasoning is through class discussions that model the kinds of reasoning we want students to do. Some teachers engaged students in debates about what a particular data pattern might mean. For example, in the *First Census* project many students noticed a pattern of the locations of the biggest American cities in 1790: that only one major city was located in the South (Charleston, SC), and most were in the Northeast (see Figure 5.4, first map). The teacher asked for possible explanations of this pattern, and one suggestion was that there might not have been many people living in Southern states in 1790. The class could check this explanation by turning on the "Total Population" layer (Figure 5.4, second map), which showed that the most populous state that year was Virginia, a Southern state.

Another possible explanation was that Southerners might have lived mainly on farms, and not in cities. This could not be directly checked in the GIS census data, but students observed the large enslaved populations in Southern states (Figure 5.4, third map), and some suggested that these people must have worked on plantations. Questions about whether enslaved people might have also lived and worked in cities led to further investigation of the data, resulting in a series of connected claims, evidence, and questions for research in the documents.

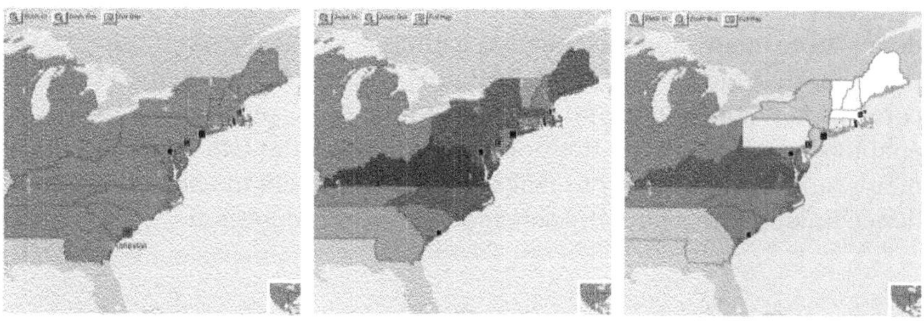

Biggest cities by population, Charleston SC selected | **Total population by state** | **Enslaved population by state**

Figure 5.4. Three GIS maps from the "First Census" project, showing three different census data variables from 1790 (population by city, population by state, enslaved population by state)

By guiding such a series of questions, observations, and possible explanations, teachers could lead these discussions toward productive themes for the investigation. In the *First Census* project, these students needed to develop an understanding of differences among Northern, Mid-Atlantic, and Southern regions around the time of the Constitutional Convention, and to reason about differences between small and large states. These understandings were important in order to engage in historical arguments about the Great Compromise, the 3/5 Compromise, and the significance of the economics of slavery in early American government and society. Another *GIS for History* unit, *Slavery in America*, provides data maps and documents for an in-depth examination of the institution of slavery, which could follow later in the year for these students.

Discussions like the one described above provided opportunities to model historical thinking by directing students to documents and census data for more evidence to support their explanations. In addition to scaffolding arguments in class discussions, some teachers had students gather information from the GIS and the documents using a structured note-taking form. An electronic "evidence worksheet" (a Word document saved and edited by each student) provided spaces to copy and paste evidence and describe its significance, and to organize this evidence to answer questions about the historical topic.

For example, the Great Migration evidence worksheet (Figure 5.5) asks students to investigate three broad questions:

1. Where did African Americans leave from, and what were their destinations?
2. Why did this migration happen?
3. What helped migrants to move?

Under each question students copied and pasted several pieces of evidence from maps and documents, gathered as they worked, and elaborated on what each one showed. This note-taking structure helped students connect key concepts from the curriculum (e.g., push and pull forces, enablers of migration) to their observations from the GIS maps and documents, and directed them to make specific points about each piece of evidence linking it to a larger observation about the migration.

Connecting Data Maps to Human Perspectives

Although GIS is primarily a quantitative data analysis tool, students need to use it not just for statistical reasoning, but for historical thinking about human actions, experiences and motivations. *GIS for History* supports this

Name: Period: Date:

The Great Migration 1900-1960

1. Where did African Americans leave from, and what were their destinations?

The African Americans left from the South and spread all over the country. They were now not only in the South, they were in the North, East and West.

EVIDENCE: Paste evidence from letters & maps	**ELABORATION:** What does this evidence show?
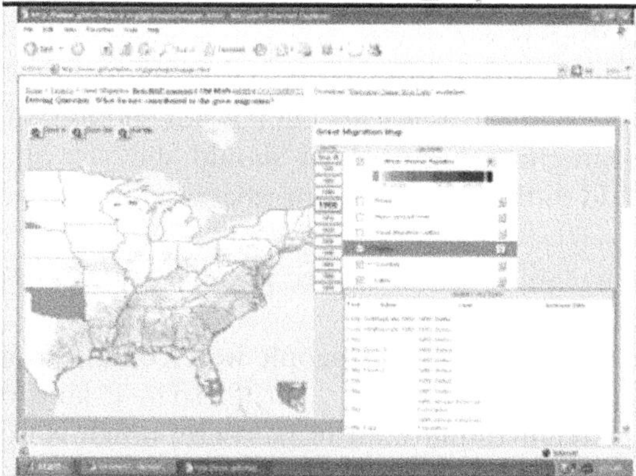	This is a map during1900; it shows very low population of African Americans in the North. There is higher number of African Americans in the South than any where else in the U.S.
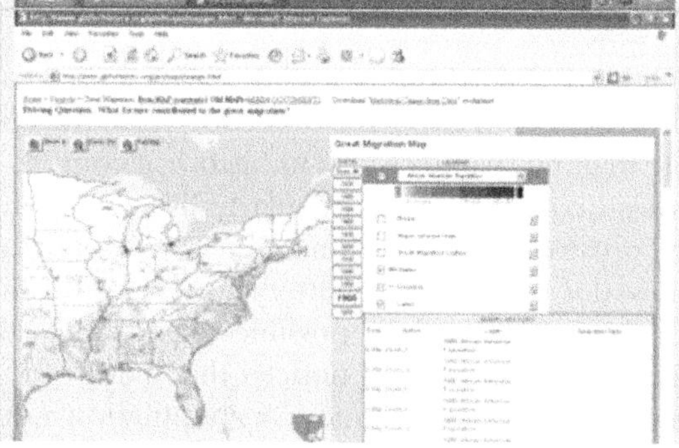	This is a map during 1960; it shows African Americans not only in the South but also in the North. Most of them living in Cook County, IL. and Wayne County, MI.

Figure 5.5. Part of a students "evidence worksheet" from the Great Migration project. Students write an answer to each question, provide evidence supporting each answer (copied and pasted from the GIS and the on-line documents), and elaborate on what each piece of evidence shows.

by combining document analysis with the geographic and mathematical observations made with the GIS maps. We believe that GIS can contribute to meaningful reasoning about historical actors, helping students place them in geographic and demographic context, and using data to generate insightful questions about people and the contexts surrounding historical events. Teachers can help students connect their observations of data patterns with thinking about real people in particular times and places. For example, after an initial class discussion about a GIS map showing African American population data, a student here moves the discussion of a data pattern to an insight about the actual people represented by the data. An objection has just been raised to the claim that "African Americans left the South" during the Great Migration, given the large numbers living in the South in 1960:

> Teacher: [Quoting a statement by a student] "African Americans left the South and moved to the North in the early 1900s." True or false?
>
> [*Several students respond at once*]
>
> Teacher: True and false, right? This is true, but it is not enough, she has to say something more. What should she say, what should she add to this to account for what we just saw?
>
> [*Students suggest ways to re-phrase the claim*]
>
> Teacher: So she can say, "Still many remained in the South," is that good? ...
>
> Student: No because it kind of says that people were coming from the South and some just stayed ... It doesn't—it kind of tells you they really didn't want to be there, which is not necessarily true either.

The student's objection ("it kind of tells you they really didn't want to be there, which is not necessarily true") is an explicit connection from the colors on the map to real people making real decisions in historical context.

Teachers can reinforce these connections, particularly by connecting historical documents to the contextual information in the GIS about the places and times in which those documents were written. For example, one teacher was helping his students think about the possible role of railroads as enablers of the Great Migration. Students had noticed that many of the counties with the largest increases of African American population were located along major railroad lines, such as the Illinois Central. While some students considered this sufficient evidence that the railroads were enablers, the teacher encouraged those who questioned this evidence:

> Teacher: "The railroad helped African Americans to move north." True or false? ... A lot of you guys put down railroads for enablers. ... [earlier a student] raised her hand and she said, "How do I know the railroad is an enabler?" And she said, "Were they allowed to ride the railroad?"

He went on to ask students to answer this question, which cannot be answered using the GIS: were African American Southerners able to use railroads as a means of transportation in the early 20th century, or were they prevented by Jim Crow laws? The answer to this question is key for drawing appropriate inferences from the GIS map.

Several students remembered a photograph they had seen showing African Americans waiting for a train in a Florida railroad station in 1921 (shown in Figure 5.6). During this discussion some students looked through one of the document archives linked from the website (http://www.inmotionaame.org/gallery/), and found other photos showing migrants journeying north on trains. This discussion helped students combine multiple sources of evidence, and connect the data patterns on the map to real people living their lives in the historical period being studied (see Figure 5.6).

EVIDENCE of enablers: Paste evidence from letters & maps	**ELABORATION:** What does this evidence show?
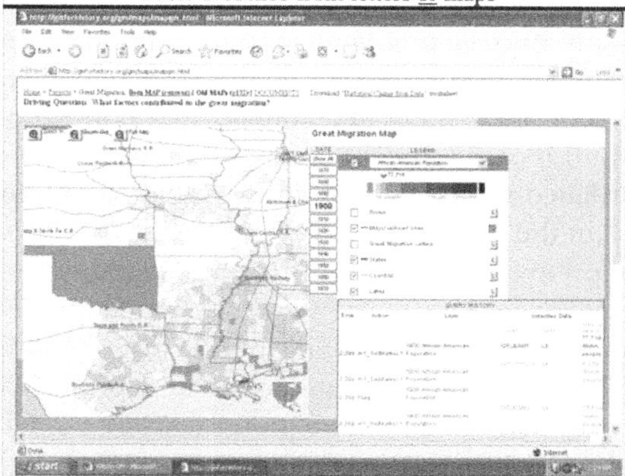	This show that the Illinois central R.R ran from Orleans to Cook in the 1900's, which helped as good transportation for the African Americans, which would be known as an enabler
	This picture show African Americans waiting at a train station so the they can go to their destination sometime in the 1900's

Figure 5.6. A student's evidence worksheet combining a GIS map labeling major railroad lines (ca. 1900) with a photograph of a Tallahassee segregated railroad station waiting room.

Reasoning Fallacies and GIS

As in all inquiry, students must learn not to over-generalize; to take into account their sources when interpreting data; and to differentiate inferences from observations, and claims from evidence (Kuhn, 1991). GIS data maps can easily be misused in these reasoning processes. A common mistake is the "ecological fallacy" (Robinson, 1950): attributing observations about a geographic area (e.g. aggregated demographic or political characteristics of a whole region, state or county) to all inhabitants or places within that area. An example of this fallacy from the Great Migration GIS investigation would be a student claiming that all African American inhabitants of Hinds County, Mississippi, wanted to leave the South, citing as evidence a decrease in the African American population of that county. The trend in aggregate data does not support claims about any particular town or individual within that county. Equally problematic is the converse: overgeneralizing from specific information found in a historical document (e.g., evidence of motivations of one person in a county) to a population (all people in that county). In this case a student might read a letter from a hopeful migrant seeking unskilled industrial work in the North, and infer that all migrants lacked skills in the trades.

Perhaps the most common reasoning error (which is certainly not unique to GIS!) is equating correlation with causality. The student quoted in the railroad example above showed an awareness of this distinction in her question: correlation of railroad lines with in-migration counties did not prove that the railroad played a role in that migration. While the GIS maps are designed to reveal important relationships, not all relationships in the map are meaningful. Students may "fish around" for any information in the map, and assume that it must explain the phenomenon being studied.

While students may easily misuse GIS maps in these ways, it is important to note that students need opportunities to make these mistakes and receive a teacher's feedback. This kind of practice in historical reasoning from evidence is the only way for them to develop these skills. GIS investigations give teachers a chance to engage in specific discussions of these reasoning fallacies, to assess students' use of historical evidence, and to give them practice in constructing arguments from evidence.

CONCLUSIONS AND FUTURE DIRECTIONS

GIS for History was designed to facilitate specific kinds of historical reasoning about specific themes, time periods, and topics. Our pilot studies in high school history classrooms suggest that students often make insightful observations, attempt historical explanations, and raise interesting questions about the past when they have opportunities to explore historical data in a GIS. While it may introduce challenges and complexity, the GIS can also help teachers place critical historical questions and themes on center stage in students' minds.

GIS for History is designed for accessibility, to fit easily into the complex logistics of everyday classroom activity. No software need be installed, making it easy for teachers to take their initial step into using GIS with their students through a website. Data are accessed with a single click, and reported within the same interface as the map, so that observations can be made and shared easily, without opening separate reports, data tables, or windows. The Query History provides a simple trail, to make it easy for students and teachers to retrace their footsteps in an investigation. Users do not have tools to change the visual representation scheme, to add or remove layers, or to create their own maps, keeping the interface simple and the data representations consistent.

These decisions in designing the site were meant to support teachers in quickly accessing relevant materials, and to scaffold historical reasoning for their students. Many of the low-level logistical burdens of creating the GIS maps were removed. We believe that, with more experience, some teachers—and many students—are likely to be interested in eventually going beyond the limitations of the *GIS for History* investigations. In collaboration with Daniel Edelson, director of Northwestern University's GEODE project and designer of *My World* GIS software (see Edelson, Smith, & Brown, this volume), project files are in development for each curriculum unit in *GIS for History*. These files will be available for download and use in *My World*, enabling students to do more complex queries and other more labor-intensive work with GIS.

As the community of GIS-savvy history teachers grows, we envision a continuum of GIS integration in classrooms, to build students' observational skills and representational fluency over time. Students may begin with web-based, structured investigations with tools like *GIS for History*, and move up in complexity toward more ambitious projects using professional GIS tools. Dimensions of increasing complexity could include those listed in Table 5.2.

While the prospect of these extensions toward more complex historical GIS inquiry is exciting, we also believe that there is a meaningful place for the kinds of investigations described here using *GIS for History*. They are not just a means toward an end result of more advanced GIS use by all students and teachers. The power of the simple data browsing tools, designed for tight integration with document analysis within history curriculum, is a significant step forward in enabling teachers and students to engage meaningfully with complex historical data.

TABLE 5.2
Dimensions of Complexity in GIS Investigations

Dimension	GIS for History	MyWorld	Professional GIS tools
Control over map layers	Display/hide pre-selected layers	Add/remove layers, import new layers	Create new layers, assemble original GIS
Control over representational schemes	Interpret schemes using legend	Choose from a library of schemes, modify schemes	Determine schemes
Query tools	Click map to get data values	Selection tools, mathematical and logical query tools, save query results	Advanced query tools, generate new layers and maps from query results
Scope of investigation	Guided inquiry on pre-selected topics	Support for guided investigation, options of using project files, or open-ended use	No supports for guided investigation, open-ended GIS use

ACKNOWLEDGEMENTS

This material is based on work supported by a "We the People" grant from the National Endowment for the Humanities (NEH), Grant # EE 50023-04, and also by the National Science Foundation (NSF) under Grant # 0337598. Any opinions, findings, conclusions, or recommendations expressed in this material are those of the author, and do not necessarily reflect those of the National Endowment for the Humanities, nor of the National Science Foundation.

REFERENCES

Bol, P. K. (2004, March). *Putting history in geography: The challenges of creating a GIS for the history of China.* Paper presented at the Newberry Library conference History and Geography: Assessing the role of geographical information in historical scholarship, Chicago, IL.

Cunfer, G. (2002). Causes of the Dust Bowl. In A. K. Knowles (Ed.), *Past time, past place: GIS for history* (pp. 93–104). Redlands, CA: ESRI Press

Cunfer, G. (2005). *On the Great Plains: Agriculture and environment.* College Station: Texas A&M University Press.

Healy, R. (2006, November). *Railroads, regional investment and 19th century industrial development of the northeastern United States.* Paper presented at the Social Science History Association conference, Minneapolis, MN.

Knowles, A. K. (Ed.). (2002). *Past time, past place: GIS for history.* Redlands, CA: ESRI Press.

Knowles, A. K., Farrell, L., & Roush, W. (2004, March). *Visualizing Gettysburg.* Paper presented at the Newberry Library conference History and Geography: Assessing the role of geographical information in historical scholarship, Chicago, IL.

Kuhn, D. (1991). *The skills of argument.* Cambridge: Cambridge University Press.

Radinsky, J., Oliva, S., Allende-Pellot, F., Liceaga, I., Sacay, R., Singer, M., et al. (2006, April). *Quiet students in small-group work: Emergence of student roles as data investigators.* Paper presented at the Annual Conference of the American Educational Research Association (AERA), San Francisco, CA.

Robinson, W. S. (1950). Ecological correlations and the behavior of individuals. *American Sociological Review, 15,* 351–357.

Singer, M., Radinsky, J., & S. Goldman (in press). The role of gesture in meaning construction. *Discourse Processes.*

Young, J. R. (2006). With digital maps, historians chart new way into the past. *Chronicle of Higher Education, 53*(12), A33–A39.

CHAPTER 6

THE WORLD AT THE STUDENT'S FINGERTIPS

Internet-Based GIS Education Opportunities

Joseph J. Kerski
Environmental Systems Research Institute

Despite the changes in GIS in education over the past 20 years, its model of delivery has held remarkably constant over much of that time. While it is true that more spatial data has become available and accessible, and while hardware, software, and computer networks certainly became more powerful and mobile throughout the period, the hardware-software model of GIS use was much the same. This model consisted of loading and running GIS as a desktop software application on a computer running the Unix, Windows, or Macintosh operating system. The model included the loading and use of spatial data from an outside source. At first, the source of this spatial data was from physical media—magnetic tapes, ZIP drives, CD-ROMs, DVDs, or some other external device, and later became increasingly downloaded from the Internet. However, the result was still the same—the data was stored on a hard disk. The hard disk was either local, externally attached to a local computer, or on a local network.

The above model provided the foundation for 20 years of teaching *with* GIS and *about* GIS (Sui, 1995). However, since the advent of Internet-based GIS, teaching with GIS and teaching about GIS have both been slowly but

Digital Geography: Geospatial Technologies in the Social Studies Classroom, pages 119–134
Copyright © 2008 by Information Age Publishing
All rights of reproduction in any form reserved.

inexorably changing. This chapter will focus on the changes that Internet-based GIS brings to teaching *with* GIS. These changes include opportunities as well as challenges.

Maps on the Internet were one of the first things hosted as the World Wide Web was developing, but these "web-based mapping tools" were by and large static maps that the student could simply view or else manipulate online within structured limits. By the end of the 1990s, however, web-mapping tools became more powerful, allowing the user to create and interact with them. Just as in non-educational settings (Harder, 1998), Internet-based mapping applications are changing the way that people use and manage geographic information. Indeed, Internet-based GIS changes more than simply the way that spatial data is accessed and analyzed—it can and has changed some of the fundamental pedagogical models of teaching with GIS.

Web-based mapping services provide maps through Internet Map Servers (IMS), usually as rasters (images), but increasingly as vectors—points, lines, and polygons. One standard specification for such a server is the Open Geospatial Consortium's Web Map Services (WMS). One type of server is ArcIMS (Arc Internet Map Server) produced by ESRI (East et al., 2001). Interactive maps served with ArcIMS, WMS, and their open source counterparts contain maps comprised of individual layers that can be turned on and off. Each vector layer contains features that have attributes stored in tables that can be queried, selected, sorted, and investigated further. Indeed, the term *Internet-based GIS* is a term that seems to convey the sense that these web-based mapping tools and services do much more than simply provide online maps for viewing. Educators seeking to foster spatial thinking in students are increasingly using these tools alone and in conjunction with desktop GIS.

However, in one respect, today's Internet-based GIS are similar to the web-based maps of the mid-1990s: The only requirements for accessing them are a web browser and an Internet connection. The GIS database is maintained on the server side, which the user accesses but another individual or organization has built. On the user's side (the "client side"), the map viewers can be thin clients, such as a web browser, a customized client, or a desktop application such as a GIS software package. The software could range from ArcGIS (a full-featured GIS software package built for the professional GIS community but also used by educators), ArcExplorer (a free Macintosh and Windows-compatible GIS), or ArcPad (GIS software that can be loaded onto handheld computers).

Not only have Internet developments enabled students to use a web browser for spatial analysis, but the developments have affected the standard desktop GIS model as well. Inside a desktop GIS, students can add content that is not resident on their local computers, but rather, content that resides on a web server in a faraway place. The data does not have to be stored locally, but the remote data can be combined with the student's own data, or with other map layers stored on the local Intranet. This radical shift enables dynamic content to be used in the curriculum, reducing the need for educators to have to store their own data.

Web-based mapping services have evolved to the point where the user can identify features, spot street addresses on a map ("geocode" addresses), hyperlink objects such as photographs, movies, text, and URLs to mapped data, change the map projection, query maps and attribute tables, create graphs from the geographic information, and perform other functions formerly only within the realm of desktop GIS. Another recent development is that web mapping services have begun to accept a growing amount of user-generated data. These user-generated data include addresses that can be geocoded, GPS-generated coordinates for mapping, and photographs to be hyperlinked to points on a web map. Having students work with data that they collected themselves is a powerful attraction for many educators who seek to go beyond simply looking at a map to incorporate field work, the geographic inquiry process, and the scientific method in their classrooms.

These new tools present new opportunities for the social studies educator as well as new challenges. As never before, educators have a wealth of geospatial tools at their fingertips, most of which require fewer hours of training than traditional GIS. These tools may hasten the ability of educators to meet spatial learning challenges as identified in the National Research Council's *Learning to Think Spatially* report (2006). They can support standards-based inquiry-driven methods of teaching and learning, while providing basic analysis tools for exploring geographic or scientific data (Baker, 2005). Challenges in using Internet-based GIS in education include data quality, functionality, computer security, Internet bandwidth, and distractions due to the visual appeal of these new tools. These tools can be used in a variety of ways such as in teaching about population characteristics, migration, economics, and history. They can be used at any scale—from analyzing locally collected data in the field to global issues and problems. In each application in education, students can learn about the processes that are active on our planet, and the reasons why those processes exist.

INTERNET-BASED GIS SERVICES

Online GIS services are produced by universities, local, regional, national, and international government agencies, nonprofit organizations, professional societies, and private companies. They are changing rapidly and the services described below will no doubt rapidly expand in the near future.

They are incredibly varied in their scale, coverage, themes, and purpose, and are therefore difficult to classify. Nevertheless, the resources below have been successfully tested in a variety of classrooms by the author and seem to fit logically into these categories.

Mapping People at Places

One category of Internet-based GIS services might be aptly entitled "Mapping People at Places," and include services that allows for customized thematic mapping of variables. American FactFinder provides for demographic and housing variables by neighborhood from the 2000 U.S. Census of Population and Housing (http://factfinder.census.gov). Social Explorer, from Queens College of the City University of New York (http://socialexplorer.com) allows for the analysis of detailed historical U.S. Census maps and reports back to 1940. The Modern Language Association's web mapping tool allows languages spoken in the USA to be analyzed by state and by county (http://www.mla.org/census_map). Cartograms, maps that convert areas into sizes based on numeric variables while maintaining some degree of geographic accuracy, can be explored for hundreds of variables from poverty to fertility by country using World Mapper (http://www.worldmapper.org). National Geographic's MapMachine allows for mapping at different scales and dozens of themes ranging from population density, ecoregions, satellite imagery, to the number of days each year in which snow covers the ground (http://nationalgeographic.com/maps). NationMaster is a service that provides a wealth of raw and ranked statistics by country displayed as tables and graphs (http://www.nationmaster.com), allowing for import into a GIS package, and by state (http://www.statemaster.com).

Mapping Your Own Places

Another category of Internet GIS services could be named "Mapping Your Own Places" because these allow for dynamic maps to be combined with the user's own data. These include a service that enables GPS coordinates to be mapped atop satellite images. An early example was GPS Visualizer (http://gpsvisualizer.com) which allows for the user's latitude-longitude coordinates from a GPS to be mapped atop a satellite image. Another service named Batch Geocode, geocodes addresses to street maps (http://www.batchgeocode.com). An increasing number of services allow for the tagging of photographs to an online map or satellite image, such as Wayfaring (http://www.wayfaring.com/), Flickr (http://www.flickr.com), and YourGMap (http://yourgmap.com). A different kind of service, built specifically for education, maps student responses to online surveys (http://kangis.org/mapping/sdm/). The surveys can be of the student's own or the teacher's own creations.

Mapping Real-Time and Near-Real-Time Data

A variety of mapping services bring in real-time and near-real-time data to students for analysis. Examples include examining the pattern of wildfire perimeters that are only a few hours old in the USA (http://www.geomac.gov) and in Australia (http://sentinel2.ga.gov.au/Sentinel/imf.jsp), earthquakes (http://earthquake.usgs.gov/eqcenter/catalogs/), and streamflow and water quality as measured by gauging stations (http://waterdata.usgs.gov/nwis/rt). Similarly, Weather Bug provides an interactive, real-time weather map (http://ims.gis.weatherbug .com/website/ArcWebExplorerDemo.htm) that is an example of an application built on ESRI's ArcWeb Services that combines ArcWeb JavaScript Application Programming Interface with an Internet Map Service on top. By examining the data on these sites in an inquiry-driven lesson, students gain a deep impression that the Earth truly is a dynamic planet: At some scales and with certain phenomena, the planet changes by the minute.

National and Global Mapping Portals

Other web-based mapping services allow the user to make maps online and download data either dynamically into a desktop GIS or via the conventional method of saving the file and adding it as a map layer. These include portals that are the result of government cooperation and initiative, such as the Geospatial One-Stop (http://gos2.geodata.gov/wps/portal/gos). Students can use the Geospatial One-Stop, for example, to examine the changes in the settlement of New Orleans for the past century, and how Hurricane Katrina impacted the landscape. Students can use the National Atlas of the USA (http://nationalatlas.gov) or the National Map (http://nationalmap.gov) to analyze over 200 variables literally from A to Z—ranging from agriculture to the spread of the invasive zebra mussels.

These are not just mapping sites, nor are they simply starting points for two components of the geographic inquiry process—acquiring geographic resources, and exploring geographic data. Rather, Internet-based GIS represent opportunities to open up the world of geographic inquiry for students. For example, when students use the National Atlas to investigate the agriculture theme and create a map of cotton production, they can be encouraged to think about the southerly pattern, and connect this pattern to what they know about climate in the southern part of the USA. Next, students can then examine cotton production versus rainfall, and discover that while cotton grows in the south, it does so in areas that receive a moderate amount of precipitation (Georgia, for example) but also in areas that receive little precipitation (central California, southern Arizona, and the Texas panhandle). The teacher can then ask, "How can cotton be grown in Arizona?" and follow this with a value-based question, perhaps the most important of all, "Should cotton be grown in Arizona?"

Essential to successful use of Internet-based GIS in all of these examples is the pivotal role of the educator who is well-acquainted with the themes of the discipline in which he or she is teaching and knows how to frame these inquiry-based questions. Thus, a geography teacher who is familiar with the process of asking geographic questions, acquiring geographic resources, exploring geographic data, analyzing geographic information, and acting upon geographic knowledge (Geography Education Standards Project, 1994), geographic inquiry take advantage of the power of Internet-based GIS to foster inquiry. The same is true of a history teacher who seeks to use these sites for historical inquiry, of a mathematics teacher, and so on. The GIS is a key component in each of the steps in geographic inquiry, not simply as a source of data.

Other portals are the results of private enterprise and universities. The Geography Network (http://www.geographynetwork.com) is a global network of geographic information, users, and providers, allowing students to access maps of themes that they choose, and allowing them to download data. Another portal, ArcGIS Online, is a family of web-based products and services for ArcGIS users, including imagery, political maps, and street maps, among much else. Finally, the Globalis Project (http://globalis.gvu.unu.edu/) is a collaborative effort between the Norwegian United Nations (UN) Association, the UN Environment Programme, and several schools and universities in Norway. Students can analyze statistics from the Human Development Report 2003, which is the first global status report discussing how the countries of the world compare in reaching the UN Millennium Development Goals. For each goal, students can choose between indicators that show the countries' positions as well as examine how each country is making progress toward or moving away from these goals.

Mapping History

Not long ago, the only spatial data online was current, or else, slightly out of date. Increasingly, many Internet GIS services offer historical content, which is of tremendous interest to social studies teachers. The Rand McNally classroom (http://randmcnallyclassroom.com) offers historical maps through the US History Atlas and the Goode's World Atlas. Student can examine how landscapes have changed by comparing historical USGS topographic maps (http://historical.maptech.com), to current editions (http://terraserver-usa.com and http://www.topozone.com). Students can do similar work but for a much longer time span via the UK Ordnance Survey's historical maps (http://www.old-maps.co.uk) site, and through the Library of Congress (http://memory.loc.gov/ammem/gmdhtml/) and the David Rumsey historical library of over 10,000 maps (http://www.davidrumsey.com). The GIS for History project (http://www.gisforhistory.org) from the University of Illinois at Chicago is supported by a "We the People" grant from the National Endowment for the Humanities and by the National Science Foundation, and moves beyond the provision of historical maps to lessons based on historical maps and data within a GIS environment (see Radinsky, this volume).

Virtual Globes

Another category of Internet GIS services are known as Virtual Globes, so named because they display spatial data on top of a three-dimensional globe on the computer. They represent a combination of a desktop application and content that streams from the Internet. This streamed data does not become resident on the student's computer, but is updated dynamically depending on the scale and geographic area selected. Virtual Globes include but are not limited to ESRI's ArcGIS Explorer (http://www.esri.com/arcgisexplorer), NASA WorldWind (http://worldwind.arc.nasa.gov) and Google Earth (http://earth.google.com). All require a desktop program, but use data streamed from the Internet. The data includes high-resolution satellite imagery that formerly could only be purchased from private vendors at high cost. It combines this imagery with elevation data to render the Earth in three dimensions. From these initial data sets, land cover, climate, points of interest, three-dimensional buildings of the student's own construction, and now, student-created spatial data can be added. For example, ArcGIS Explorer is a small desktop client that accesses ArcGIS Server, a library of spatial data online, in three dimensions. Students can use ArcGIS Explorer to access, integrate, and utilize GIS services, geographic content, and other Web services in a 3-D earth model. ArcGIS Explorer can also use local data such as shapefiles (vector GIS layers), file geodatabases, Google Earth Keyhole Markup Language (KML) files, and a variety of images including those in JPG, GeoTIFF, MrSID, IMG, and other formats. This means that student projects that were formerly relegated to be examined in a 2-dimensional GIS can now be analyzed using a 3-D visualization tool. Much easier than is possible with desktop GIS, student content with these tools can be easily shared with others across the Internet. This fits in well with the collaborative efforts that many educators are seeking.

An Example of Teaching with Internet-Based GIS

ArcWeb Explorer (http://arcwebservices.com/awx) is a web mapping service built by ESRI that effectively illustrates the power of some of these new tools. ArcWeb Explorer was built to showcase the data available through ArcWeb Services, a library of information and tools compiled by ESRI that contains over 100 different sources of spatial content such as streets, imagery, and even live data such as traffic. ArcWeb Services are a collection of tools that enables users to integrate mapping, routing, and geographic reporting into their web sites, wireless devices, or desktop applications. With ArcWeb Services, neither the content nor the software has to be resident on a student or teacher's local computers. ArcWeb Services lets users perform a variety of spatial query and proximity searches on a host of datasets provided by some of the GIS industry's leading data providers or by other GIS users. With the routes, base maps, thematic maps, and reports, the goal in industry or in education is to solve geographic problems.

Students can use ArcWeb Explorer to investigate over 100 demographic variables down to the neighborhood level, such as total population, median income, education, age, and ethnicity. They can compare urban core neighborhoods to first ring suburbs to exurban to rural, or compare their own neighborhood to those nearby or across town.

The map displayed in Figure 6.1 was produced through just three steps: First, Chicago was selected under the "find" tool, and zoomed to the desired scale. Second, the Data Map type was selected. Third, median household income was chosen for the variable. The map in Figure 6.1 required less than 1 minute to build from the time the tool was first invoked, and no software was required other than a web browser. Therefore, students can quickly generate maps like this and quickly move to the level of geographic inquiry and analysis. This ability is one of the main advantages for educators who seek to nudge students beyond *reading* maps to *understanding* maps, but more importantly, to understanding their community, region, country, and world. Through this tool, students can think about the *processes* behind the *patterns,* and investigate the *linkages and interconnectedness* of major geographic themes, such as scale, migration, climate, landforms, vegetation, land use, resources, energy, and more. They can tackle questions that go far beyond "Describe the pattern that you see" to "Why do incomes vary across neighborhoods?", "Why are incomes high just

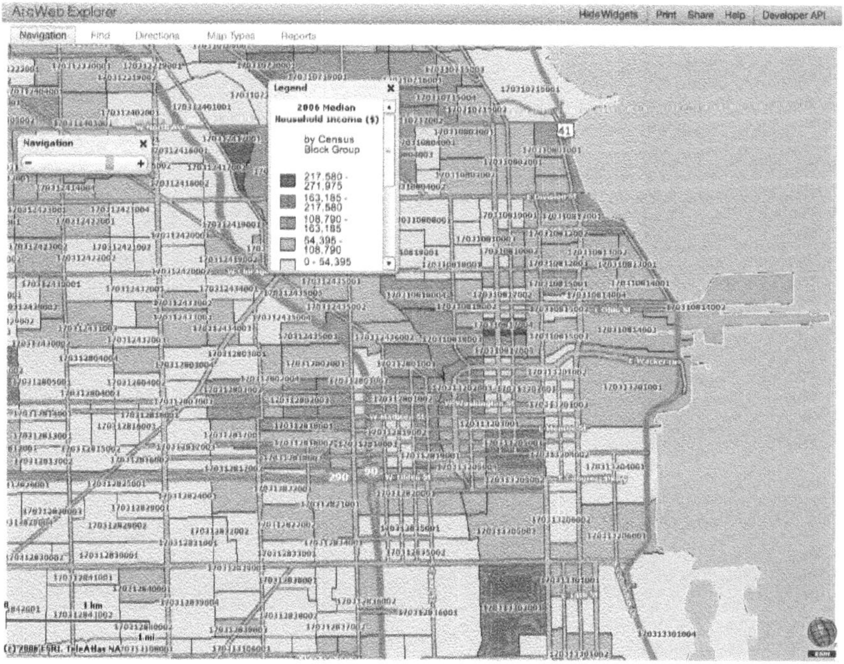

Figure 6.1. Thematic map of median income in neighborhoods in Chicago produced with ArcWeb Explorer.

outside of downtown, but decrease farther away from downtown?", and "What will happen to these neighborhoods' income levels in the future?" The advantage of Internet-based GIS allows students to more quickly get to the heart of geographic inquiry—asking the "whys of where" questions.

The same tool—ArcWeb Explorer—can be used to map the students' own data. As shown in Figure 6.2, a student created a two-column spreadsheet. The student entered the column names "name" and "phone." The student then entered the names of friends and saved the spreadsheet on the local computer. Next, inside ArcWeb Explorer, instead of finding an address, a latitude-longitude location, a computer, or a URL, the student requested the service to find an Excel spreadsheet. The student pointed the service to the spreadsheet containing the phone number listing and the service mapped them in their correct geographic locations atop a hybrid street-satellite image map. The student could then analyze the diffusion of these phone numbers, clustered on the school that they all attend, visible in the west central section of the map. They could then compare the diffusion of students that attend a neighborhood school versus the diffusion of students who attend a magnet school that attracts students from a wide variety of neighborhoods across an entire metropolitan area. They

Figure 6.2. Map of Student Telephone Numbers from an Excel spreadsheet.

could then compare the size of these "student catchment basins" for a secondary school versus a community college, a regional university, and a national and international university. They could then think about the catchment basin for schools versus businesses, and examine why gas stations have a smaller catchment area than hospitals.

The ability to map one's own data is certainly not commonplace in web mapping services, but is moving in that direction. Teachers can take advantage of these developments by having students map their field-collected latitude-longitude data with GPS receivers. They could also use online telephone directories, select certain businesses, and map them. Why do rental car companies cluster near airports, while others, such as check-cashing services, seem more scattered? These are questions that foster spatial thinking—students begin to understand that there is a reason for the location of just about everything.

BENEFITS, LIMITATIONS, AND EXPANSION OF INTERNET-BASED GIS IN THE CURRICULUM

Benefits of Internet-Based GIS in Education

The advantages of Internet-based GIS for the educator are numerous (Baker, 2005). No software is required beyond a standard web browser. This alone, for instructors who are faced with difficulties with the school technology staff in installing any software not viewed as standard in the school, is a significant benefit. Furthermore, most web-based GIS require little time to learn, drastically reducing the time it takes for educators to incorporate them into their curricula. For example, field locations that students have studied can be mapped in a few seconds from a spreadsheet within ArcWeb Explorer. The instructor does not need to find and format the base data that needs to underlie the student-collected data. Thus, the data collection and formatting are both reduced in time and complexity, leaving more time for analysis.

Dynamic content, such as real-time wildfires or earthquakes, can be analyzed online in a way not possible in the desktop environment. Online map data is often updated frequently, eliminating the need for instructors to mine the Internet for more recent data for their desktop GIS base maps. In many countries, spatial data from local and national government agencies carries significant costs, copyright, and licensing restrictions that severely limits the use of that data in schools. A significant advantage of online GIS over desktop GIS particularly in these countries is that the data are being viewed online, not downloaded, and therefore is accessible without cost or licensing restrictions.

Many web-based GIS tools can be effectively used to enhance teaching and learning in conjunction with traditional desktop-based GIS. Teachers seem better able to make the leap to using maps to *analyze data* rather than just *look at places* when there is an intermediate technology that introduces them to spatial analysis. Internet-based GIS provides that intermediate step. Using Internet-based GIS provides an excellent bridge for both teachers and students between paper maps and atlases and a full desktop GIS such as ArcGIS. It also seems to provide an excellent scaffold upon which to build spatial understanding. Typical GIS navigation such as zooming, panning, selecting, and identifying can be effectively introduced through the use of web-based mapping services. Most of these tools function in the same way within a desktop GIS environment that they do in the Internet-based environment. Furthermore, Internet-based GIS provides an excellent avenue to more advanced spatial analysis with desktop GIS. For example, if a student understands how to select the "row crops" land use from a web-based mapping service, the skill is easily transferable to a desktop GIS not simply because the tool and procedure may be exactly the same, but the process is the same.

Challenges to Internet-Based GIS in Education

Internet-based GIS does not solve every challenge to the use of GIS in the curriculum. Challenges to the use of Internet-based GIS for instruction exist just as they do in the desktop GIS environment. Most of these challenges are not easily solved with software and hardware, because they are embedded in the educational system itself (Kerski, 2003). Digital maps, whether on the Internet or as part of a desktop GIS package, present special challenges for interpretation and analysis (Cammack, 1999). For teachers who have previously relied on maps only for reference, and not for investigation, this represents an interpretation as well as a pedagogical shift.

Another major challenge is that Internet-based GIS services are still maturing. They do not contain all of the functions available in a desktop GIS, and indeed, usually contain a small fraction of the function in a desktop GIS. Furthermore, they may never contain all of the functions found in a desktop GIS software package. Therefore, students using Internet-based GIS are able to investigate fewer relationships and fewer content standards are achieved. Even fairly simple spatial queries such as "the number of 17th Century explorers that passed within 50 kilometers of the Virgin Islands" are still out of reach with most web mapping services.

For example, ArcWeb Explorer can be used to map coordinates at locations at which students have identified Googie Art (art featuring space and atomic themes that was popular during the 1950s and 1960s) in their community using a GPS receiver. However, if those same coordinates were mapped within a desktop GIS, the students would be able to add attributes to those points and make graduated symbol maps. They would be able to adjust the size of the symbols based on the age, color, style, and structures containing the Googie art that they were investigating. From these graduated symbol maps, they could then discover the spatial distribution of the art in their community. Are the points scattered throughout the community or focused on a few spots? Was the art confined to the business district? What is the age of the business district? Are there any new buildings that contain Googie art as an evidence of its renaissance? They could analyze the points in three dimensions and output them in a variety of graphs and maps as part of an oral or written report.

A recurring theme of teaching with online or desktop GIS is data quality. Maps are by definition only representations of reality. One of the first steps involved with making a map is to distort some aspect of the three-dimensional Earth onto a two-dimensional paper map or computer screen, such as area, direction, distance, or shape. On maps, some phenomena and features are included, while others are left out. Maps can be manipulated and symbolized to tell a story, and as such, are inherently value-laden (Monmonier, 1996). These characteristics do not mean that maps are not useful, but only that these limitations be kept in mind when students are using them. Much, but by no means all, of spatial data used in desktop-based GIS contains metadata—documentation that indicates who created the data, when it was created, at what scale, processing steps, and other information that helps the user to make a judgment as to the quality of the data and whether that data can be relied upon to make decisions. With Internet-based GIS, much less of the data adheres to metadata standards, and therefore, varies more widely in terms of data quality. For example, a certain web mapping application (http://www.chicagocrime.org) shows crimes in Chicago, but does the student using it know what period of time the data were collected, how the data were collected, and whether the data covers only areas within the city limits or whether it includes the suburbs?

A further challenge of web mapping services compared with desktop GIS is that because the maps and services are online, one cannot teach with them if the website or the Internet connection is down. Replication is important in education. How can a teacher teach with a tool that is working during one class period, but won't work during the next? Or, how can a teacher teach with a tool that might look completely different during the next semester or even a week later, when their lesson contains screenshots of images that are no longer valid? A back-up plan is also important in education. Teaching with Internet-based GIS is very difficult to replicate with paper maps and textbooks. One back-up plan is to use desktop GIS to approach what students are doing on the web, but this requires a teacher who is quite comfortable creating his or her own lessons with GIS. A related disadvantage for Internet-based GIS use is that, unlike desktop GIS use, it is difficult to save one's work and return to it during the next day's class period.

Furthermore, the student has less flexibility with colors and symbols with web-based services than they do with desktop GIS. Symbology is typically confined to one choice for points, lines, and polygons on the web, whereas inside a desktop GIS, the student has a nearly infinite combination of colors and symbols to use as part of the toolkit, and in addition, students can import their own custom-made symbols.

As web-based mapping tools have increased in power, many of them require a high-speed Internet connection. Although such connections are becoming more common in primary and secondary schools, they are far from ubiquitous. Often, even when a high-speed connection is available, the network experiences a sudden slowdown at certain times of the day when more classes and employees are accessing the network, or when every student is accessing a large dataset served through an Internet-based GIS.

Another disadvantage is that some of these services link to advertisements and chat rooms that either cause the entire site to be blocked by school servers or else are considered undesirable by teachers and school information technology staff. This can be overcome if the site is requested to be included on the "approved" list, but this often creates a bureaucratic hassle on a magnitude of getting software installed. Internet-based GIS are often less stable than desktop GIS, and they are often slower, since they are streaming data over the Internet rather than simply showing data on the local computer.

Finally, some web-based services, particularly visual globe software with their three-dimensional interfaces, become so captivating to many students that they have a difficult time concentrating on the problems assigned. At that point, the tool becomes entertainment, geographic entertainment to be sure, but entertainment just the same, enticing the student to flip or rotate the Earth, fly to places that are not part of the lesson, or add content that is not part of the lesson. Much has been researched about this "time off task" (Abramowitz, O'Leary, & Rosén, 1987; Bru, 2006) but the issue is compounded with computers. It is one thing to have students off task while they are using paper maps or textbooks, and quite another when the source for temptation is literally in front of their faces, on a computer monitor. A related concern is how much "control" an instructor feels is necessary for successful teaching. For many educators, using the Internet for anything besides research for term papers is venturing into uncomfortable instructional territory.

CONCLUDING THOUGHTS

As the GIS education community has evolved, so have the tools that community has used. One of the most powerful new toolkits is web-based mapping services. These have evolved to the point that they truly are an Internet-based GIS. To be sure, they have fewer capabilities than desktop GIS software, but they have enough functionality to be termed "GIS" and not simply a mapping service. Internet-based GIS will continue to develop at a rapid pace, combining functions from a variety of different services in what are known as "mashups." In web mapping and GIS, a mashup is a website or application that combines spatial content from more than one source into an integrated map, tool, or visualization. Indeed, one of the most important developments will be the ability to share spatial analysis procedures and models. Just as spatial data and projects have been shared in the past, the GIS community will be able to share the procedures they used to make decisions with—from siting a new school building to assessing the best location to study burrowing owls. These procedures, shared through a common developer model, will be able to be sent to another user, who in turn can use the same procedures but also can adjust the input parameters for their own needs. Educators will be able to take advantage of these tools for a variety of curricular goals.

Just as with desktop GIS, research is lacking regarding the use of Internet-based GIS in education. The few studies that have been conducted on the effectiveness and limitations of Internet-based GIS emphasize teaching *about* GIS, rather than teaching *with* GIS (e.g., Painho, Peixoto, Cabral, & Sena, 2001). For these technologies and methods to be adopted in more schools, more research is clearly needed (Baker & Bednarz, 2003).

In his book *Earth In Mind: On Education, Environment, and the Human Prospect*, David Orr (1994) wrote, "Now more than ever, we need people who think broadly and who understand systems, connections, patterns, and root causes...how to think in whole systems, how to find connections, how to ask big questions, and how to separate the trivial from the important" (p. 23). As today's Earth issues become increasingly complex, cross scales from local to global, and begin to affect people's everyday lives, educators and society are encouraging students and future decision-makers to think outside the confines of the software or tool in which they are working. In essence, their goal is to encourage students to think "outside the box." This "box" was formerly the confines of desktop GIS software, but even if it increasingly becomes Internet-based, the tools are just that—a box. Educators seek to build the decision-makers of tomorrow who can think about all of the relevant issues that have bearing on a problem. Each year, the planet becomes more intertwined. Each action increasingly affects other people as well as the local environment, and indeed, the entire planet. Geotechnologies are becoming an increasingly critical decision-making tool that many educators feel must be included in education if society is to solve the complex problems of today and tomorrow.

REFERENCES

Abramowitz, A., O'Leary, S., & Rosén, L. (1987). Reducing off-task behavior in the classroom: A comparison of encouragement and reprimands. *Journal of Abnormal Child Psychology, 15*(2), 153–163.

Baker, T. (2005). Internet-based GIS mapping in support of K–12 education. *The Professional Geographer, 57*(44), 44–50.

Baker, T., & Bednarz, S. (2003). Lessons learned from reviewing research in GIS education. *Journal of Geography, 102*(6), 231–233.

Bru, E. (2006, February). Factors associated with disruptive behaviour in the classroom. *Scandinavian Journal of Educational Research, 50*(1), 23–43.

Cammack, R. (1999). New map design challenges: Interactive map products for the world wide web. In W. Cartwright, M. Peterson, & G. Gartner (Eds.), *Multimedia Cartography* (pp. 155–171). New York: Springer Publishing.

East, R., Goyal, R., Haddad, A., Konovalov, A., Rosso, A., Tait, M., et al. (2001). The architecture of ArcIMS, a distributed Internet map server, in *Advances in Spatial and Temporal Databases: 7th International Symposium. Vol. 2121* (p. 387). Berlin: Springer.

Geography Education Standards Project. (1994). *Geography for life: National geography standards.* Washington, DC: National Geographic Research and Exploration.

Harder, C. (1998). *Serving maps on the Internet.* Redlands, CA: ESRI Press.

Kerski, J. (2003). The implementation and effectiveness of GIS technology and methods in secondary education. *Journal of Geography, 102*(3), 128–137.

Monmonier, M. (1996). *How to lie with maps* (2nd ed.). Chicago, IL: University of Chicago Press.

National Research Council. (2006). *Learning to think spatially: The incorporation of geographic information science across the K–12 curriculum.* Washington, DC: National Academy Press.

Orr, D. (1994). *Earth in mind: On education, environment, and the human prospect.* New York: Island Press.

Painho, M., Peixoto, M., Cabral, P., & Sena, R. (2001). WebGIS as a teaching tool. In *Proceedings of the 2001 ESRI User Conference.* Retrieved March 13, 2007, from http://gis.esri.com/library/userconf/proc01/professional/papers/pap910/p910.htm

Sui, D. (1995). A pedagogic framework to link GIS to the intellectual core of geography. *Journal of Geography, 94*(6), 578–591.

CHAPTER 7

GEOSPATIAL ONLINE LEARNING ACTIVITIES FOR MIDDLE SCHOOL STUDENTS

Sven Fuhrmann
Texas State University—San Marcos

Alan M. MacEachren
The Pennsylvania State University

Mark Gahegan
The Pennsylvania State University

Roger Downs
The Pennsylvania State University

Maps, statistics, and their mutual relationship play important, often ubiquitous, roles in our lives. Humans utilize maps to look up a variety of spatial information in the environment (restaurants, driving directions, etc.) and apply statistics for identifying patterns (e.g. crime areas) or conducting preferred location analysis (e.g. a house search). Many of the applied spatial and statistical datasets are publicly available through local, state and federal government agencies. Over the past several years, government web-based services, such as FedStats (www.fedstats.gov) or the National Atlas (www.nationalatlas.gov), have provided increasingly comprehensive reposi-

Digital Geography: Geospatial Technologies in the Social Studies Classroom, pages 135–145
Copyright © 2008 by Information Age Publishing
All rights of reproduction in any form reserved.

tories through which citizens can look up statistical and geospatial information on the Internet (see Kerski, this volume).

Research suggests that it is impractical to develop a "one-size-fits-all-ages" web portal (Large, Beheshti, & Rahman, 2002; Sullivan, Norris, Peet, & Soloway, 2000). While most governmental web portals aim to reach a wide array of possible users, many neglect the needs and abilities of young information-seeking visitors. It can be challenging for children to use websites that are designed for adults. In a usability study, websites designed for adult use achieved only a 55% success rate for young teenage users (Nielsen, 2005). Nielsen (2005) concludes that the poor website performance might be caused by: a) insufficient reading skills, b) less sophisticated research strategies, and c) lower patience levels (in comparison to adults). Poor website usability and usefulness often lead to user frustration and abandonment of the information source, especially if technological difficulties are paired with limited domain literacy (Arseneault, 2003).

Federal, state and local government agencies need to provide more age appropriate web content to the growing number of young web users. In addition to redesigning existing websites and making them more accessible to students, special web portals for young visitors need to be created. The GeoVISTA Center at The Pennsylvania State University developed such a web portal (named *MapStats for Kids*) and a set of five geospatial online learning activities for the federal government statistics website (www.fedstats.gov). The online learning activities teach middle school students how to use statistical summaries and their graphic representations for a range of special topics. The content of the learning activities is closely related to the National Geography Standards and the Principles and Standards for School Mathematics. All online learning activities and the web portal were tested for usability before their release. Teachers, media specialists, parents and students were part of the evaluation process for these novel and age appropriate learning activities. The *MapStats for Kids* web portal on the FedStats website receives an average of 300 visitors per day (user statistics determined by Information Technology Services at the U.S. Census Bureau).

THE CONCEPTUAL APPROACH OF *MAPSTATS FOR KIDS*

The foundation for this research was the NSF Digital Government Project "Quality Graphics for Federal Statistical Summaries" (dgQG). Academic researchers and federal agency personnel collaborating on this project identified a series of public information access and dissemination challenges to which project research activities were directed. One of these challenges was to make a federal website more accessible to children and more useful for promoting learning related to the wide range of activities undertaken by the federal government.

The research was conducted in cooperation with the FedStats data dissemination program. FedStats is a cross-agency Internet portal designed to help visitors find statistical facts for making personal or business decisions, conducting research, and becoming better informed for participation in civic affairs and public policy debates. The goal of the *MapStats for Kids* project was to provide learning materials that will help prepare school students to take advantage of the kinds of information provided through FedStats. Specifically, the goal was to design a web portal that met three core objectives:

1. provide online learning activities for middle school students within the FedStats Internet portal,
2. interest students in statistical information (and the maps and graphics used to communicate or work with that information), and
3. help students learn how and where to use such tools.

The *MapStats for Kids* online learning activities focus on geospatially-referenced, federal statistical summaries and match the geospatial and statistical knowledge skills, interests and experiences of middle school students with the current national education curricula, especially the "Principles and Standards for School Mathematics" (National Council of Teachers of Mathematics, 2000) and the "National Geography Standards" (Geography Education Standards Project, 1994).

Web-based interactive maps and graphic-based learning activities offer possibilities for visualizing federal statistics and at the same time can support an integrated approach to educational standards across the subject areas. Our approach in the *MapStats for Kids* project was built on two fundamental ideas: a) fostering an exploratory, flexible, problem-solving approach to the use of statistical data, and b) supporting the development of sets of knowledge and cognitive skills that can underpin the use of statistical data. For middle school students, as for many people, statistical data are most useful when they can be readily converted into information that students can apply to answering a question or solving a problem. Useful and usable visual displays and interactive visualizations of quantitative information are a key to data exploration, problem solving, and the development of knowledge and cognitive skills (Tufte, 1983).

A core objective for *MapStats for Kids* was to attract middle school students by providing interesting, accessible, challenging, informative and useful content. Crucial to this idea was: a) an appreciation of the cognitive skills that underpin successful performance (Liben, 1999); b) a recognition that the skills develop over time as children mature (DeLoache, 1989);

and c) an awareness that there are individual differences in performance of the same tasks for children at the same age range (Liben, 2000). While children need to develop many abilities (e.g., logico-mathematical, linguistic, music, spatial, bodily-kinesthetic, interpersonal and intrapersonal skills) (Gardner, 1998), the goal of our geospatial online learning activities was to support the development of three skills that are central to statistical and geospatial data analysis: logico-mathematical, representational, and spatial skills.

LEARNING ACTIVITIES: FROM THEORY TO PRACTICE

During the development of the *MapStats for Kids* web portal and online learning activities, team members from the GeoVISTA Center at The Pennsylvania State University collaborated with the Bureau of Labor Statistics (BLS) to conduct a variety of usability assessments such as an online-questionnaire, focus group test, and expert review with media specialists and teachers. The information gathered was used in an iterative revision process that was essential for the successful publication of *MapStats for Kids.*

The close interaction between teachers, media specialists, and the GeoVISTA research team allowed for quick and successful improvements related to matching the skill development requirements of middle school students and assessing technical flaws in the online learning activities and the web portal. For example, a focus group test was conducted with five teachers and media specialists in the media center of a high school in Maryland. It was the first time that the online learning activities were "officially" tested in a school environment by the research development team. During this focus group (and following testing sessions) "theory met practice." At the time of the focus group testing in 2003, the high school featured computers with Pentium II, 300 MHz processors, 128 Megabyte RAM, 800 × 600 pixel graphics cards and the Windows 95 operating system. Although everyone involved in the focus group session believed that this media center could be considered a good computer facility, it was below the standard expected by the development team. Working with high-speed computers, the geospatial application development team had already downgraded the system configuration for the online learning activities, but this and other "reality checks" required a change in the minimum system requirements to provide accessibility to everyone. Since many classrooms feature older computers and many families still connect to the Internet through dial up 56k modems, the web portal was designed with low performance computers and networks in mind. The online learning activities require Macromedia (now Adobe) Flash Player 7 and a minimal system configuration of a 266 MHz processor, 64 Megabyte RAM, and a monitor that supports the display of 256 colors at 800 x 600 pixel resolution. This minimal system configuration was tested on various operating systems before it was published on the FedStats website.

The text-based information representation components of the web portal contain mainly HTML content and simple graphics. The online learning activities and some graphic components of the web portal were designed using Macromedia's (now Adobe's) Flash software. Flash is a multimedia authoring environment that allows the distribution of program code, vector and bitmap graphics, and streaming audio and video for learning applications. Flash allows designers to quickly develop dynamic applications that are visually appealing, highly interactive, web compatible, and highly usable (if correctly designed). Flash applications run well on older computers and the software required to view Flash content is bundled with most web browsers or is otherwise free from the Adobe website (http://www.adobe.com). One important disadvantage to Flash is that the online learning applications do not fully fulfill the Rehabilitation Act (Section 508) requirements to provide web accessibility for users with disabilities (Theofanos & Redish, 2003). Adobe has started to address these accessibility concerns and has integrated features for disabled users into the most recent Flash version. Overall, Flash has become the de-facto standard for dynamic web content development and it provided all functionalities for the design of our geospatial online learning activities.

The Web Portal and Learning Activities

The design of the *MapStats for Kids* web portal and the online learning activities (http://www.fedstats.gov/kids/mapstats/index.html) was guided by three human factors: reading skills, research strategies, and patience level. As part of our efforts to ensure age and skill appropriateness of the texts provided and the compliance with national educational standards, the website and the learning activities were proofread by a teacher. The web portal itself provides quick and easy access to the online learning activities and additional resources such as help, concept explanations, and information for teachers and parents. Two characters, "Stixie" and "Globie" guide users through the website, explaining the statistical and mapping concepts used in the learning applications, and encouraging students to learn about mapping and statistics. The information resources for parents and teachers describe the technical requirements, learning objectives, educational benefits, and national educational standards that are associated with each of the learning activities. The *MapStats for Kids* web portal also provides links to the Principles and Standards for School Mathematics and the National Geography Standards (Figure 7.1).

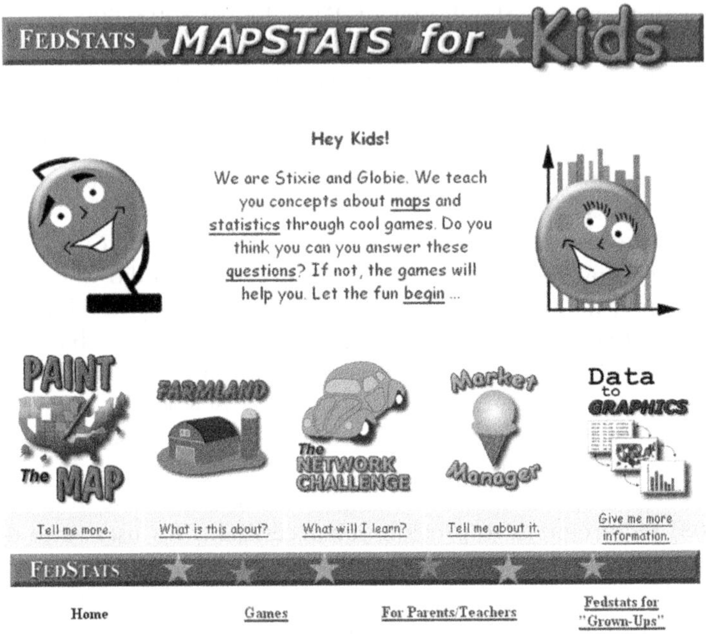

Figure 7.1. The *MapStats for Kids* web portal.

The five learning activities are called: a) *Paint the Map*; b) *Data to Graphics*; c) *The Network Challenge*; d) *Market Manager*; and e) *Farmland*. *Paint the Map* is a map-puzzle in which users build (or paint) a map to represent state level statistics. *Paint the Map* introduces the statistical concepts of mean and median to the students and allows the comparison of U.S. states on a variety of factors, such as population change, household income or energy consumption. Students decide if a certain U.S. state falls above, below, or equal to the specified measure of central tendency, then identify the state on the map and "paint" it with the color that represents the value of the statistical factor (Figure 7.2).

Data to Graphics supports the ability to move between graphical and tabular data representations for data analysis. Students learn about ranking, value categories, central tendencies, and value representation through bar graphs. The activity fosters an understanding of typical summary statistics through tabular and graphical representations by exploring the connections among a spreadsheet, a map and a bar graph. The activity represents state-level statistics, such as energy consumption and population data, in a table. Students need to find the value for a U.S. state in the table and adjust the corresponding bar graph for that state so that the height of the bar represents the same value in the table. As in *Paint the Map* the students need to color a choropleth map by identifying the state and its correct value. Stu-

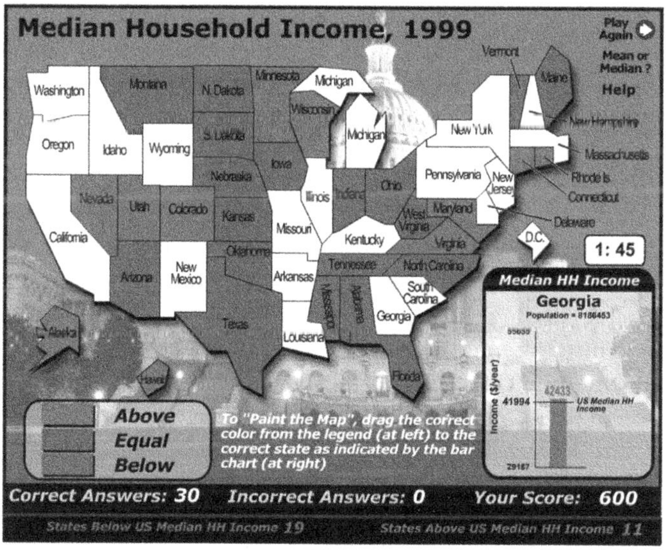

Figure 7.2. Coloring U.S. states in *Paint the Map.*

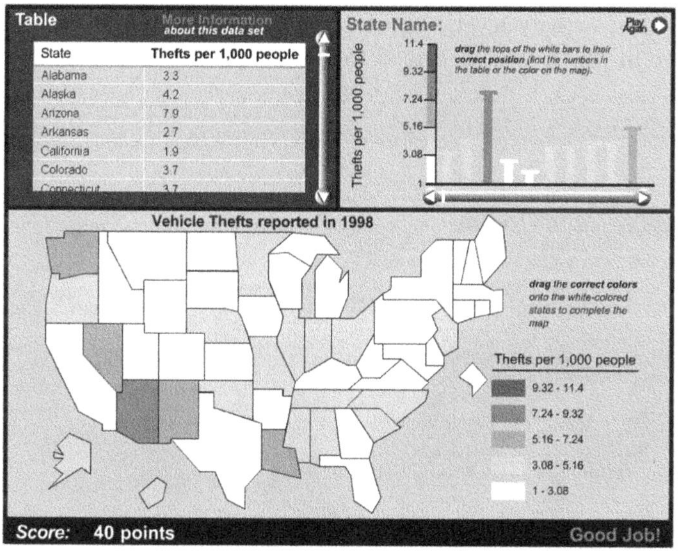

Figure 7.3. The Data to Graphics learning activity.

dents can find the values for the choropleth map either in the bar graph or the table. This application will help students to understand that statistical data can be represented in many different ways (Figure 7.3).

The Network Challenge focuses on the concept of networks and addresses the understanding of topological relationships and problem solving strategies to support spatial decision-making related to traversing a network. The student has to plan effective routes for vacation trips, band tours or political campaign stops using demographic, climatologic and/or elevation data. Each activity starts at a randomly selected state capital and the student can utilize tables, bar graphs, look-up cards and maps as information sources. While moving along the street network between U.S. state capitols, students need to keep the overarching goal in mind, e.g. reaching as many fans as possible, driving through mostly sunny states, and so on. *The Network Challenge* requires one not to only make the "next best move", but to think about how one choice relates to the next decision by utilizing different data representations (Figure 7.4).

The Market Manager focuses on the concept of regions and provides region-building exercises for three different scales: national, state and greater metro area. At each scale, the goal is to classify the units (states, counties, or ZIP codes) into four continuous regions (market territories) with equal aggregate values (e.g., populations of children aged 5–17 in each territory). Besides different scales *The Market Manager* also provides two different activity levels. Level 1 contains the ice cream distribution challenge in which students need to create four comparable sales regions

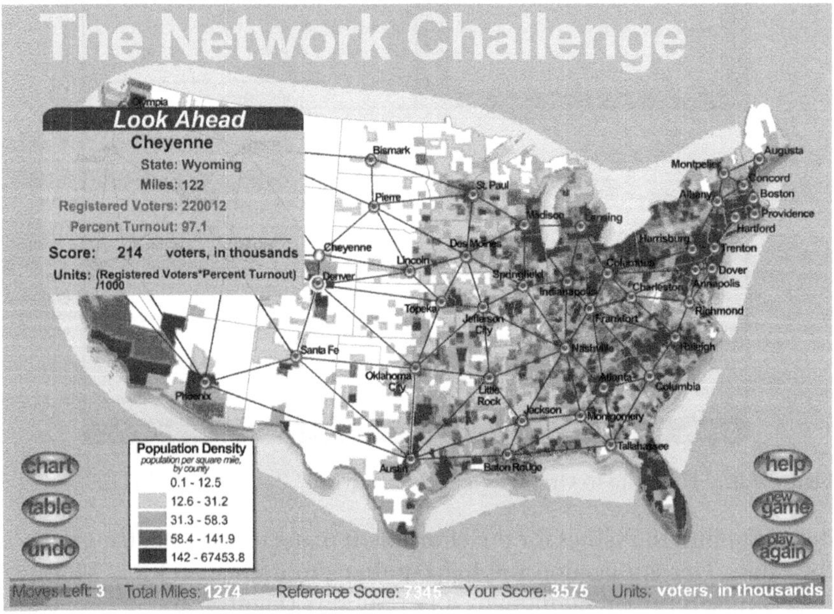

Figure 7.4. Route planning with the Network Challenge.

in which they want to distribute ice cream for free to their peers. Level 2 allows educators to customize the application and select different datasets and representations (within predetermined constraints).

Farmland teaches about changes over time in statistical datasets and shows these changes through temporal animations. The learning activity displays three different types of farm data by U.S. state: number of farms (1910–1999), average farm size (1950–1999), and total farmland (1950–1999). *Farmland* provides a video player-based graphical interface, which supports playing the animation, stepping through the animation, pausing and resetting the display. Students learn about how choropleth maps can depict multiple ordered categories of numerical data with color symbolization and how map animations are an effective way of displaying temporal changes in statistical datasets.

CONCLUSIONS

Designing a usable and useful web portal and geospatial learning activities for school children is not an easy task, especially when the design has to be compatible with a U.S. Federal Agency website design. The chosen design approach was essential in defining requirements for the online learning activities and testing the developed prototypes before publishing them on the FedStats website. For example, the focus group with teachers and media experts was guided by three overarching questions: a) Do the online learning activities appropriately support school curricula? How should they be redesigned, if not? b) Are the user interfaces generally suited for the particular age range (usability)? and c) Can middle school children use the applications to understand the data represented (usefulness)? We found in this study that the online learning activities met the middle school curricula and the participating teachers indicated that they could imagine using the same tools at the elementary and the high school levels. The participants also indicated that the contents of the learning activities are challenging enough and that the applications could be implemented for individual, group work or home work. Overall, the study revealed that the *MapStats for Kids* content

1. is likely to attract middle school students' attention and interest, thus encouraging them to explore the web portal and activities,
2. includes stimuli that are viewed as fun, easy, and interesting to use, and
3. provides a positive learning experience.

MapStats for Kids receives an average of 300 visitors per day and illuminates the need to develop age appropriate web content for young information-seeking visitors on federal, state and local government websites. In future projects and collaborations, we will continue enhancing the usefulness and usability of web portals for children. Please feel free to visit *MapStats for Kids* at http://www.fedstats.gov/kids/mapstats/index.html to experience the learning activities.

ACKNOWLEDGEMENTS

The work reported here was supported by a grant from the National Science Foundation (#EIA9983451) and by funding from the FedStats Taskforce. Any opinions, findings and conclusions or recommendations expressed in this material are those of the author(s) and do not necessarily reflect the views of the funding agencies. We wish to also thank in alphabetical order Tanuka Bhowmick, John Bosley, Stephen Crawford, Marshall DeBerry, Mark Harrower, David Howard, Rachael LaPorte Taylor, and Bonan Li for their inputs to and/or advice on various stages of the work reported here.

REFERENCES

Arseneault, C. (2003, March). *Having fun or finding information? Usability for kids sections of Web sites.* Paper presented at the 7th Annual Museums and the Web Conference, Charlotte, North Carolina, USA.

DeLoache, J. S. (1989). The development of representation in young children. In H. W. Reece (Ed.), *Advances in child development and behavior* (pp. 1–39). New York: Academic Press.

Gardner, H. (1998). A multiplicity of intelligences. *Scientific American, 9*(4), 18–23.

Geography Education Standards Project. (1994). *Geography for life - National Geography Standards.* Washington: National Geographic Research & Exploration.

Large, A., Beheshti, J., & Rahman, T. (2002). Design criteria for children's web portals: The users speak out. *Journal of the American Society for Information Science and Technology, 53*(2), 79–94.

Liben, L. S. (1999). Developing an understanding of external spatial representations. In I. E. Sigel (Ed.), *Development of mental representations - Theories and applications* (pp. 297–321). Mahwah: Lawrence Erlbaum Associates.

Liben, L. S. (2000). Map use and the development of spatial cognition: Seeing the bigger picture. *Developmental Science, 3,* 270–274.

National Council of Teachers of Mathematics. (2000). *Principles and standards for school mathematics.* Retrieved February 26, 2003, from http://standards.nctm.org/document/index.htm

Nielsen, J. (2005). *Usability of websites for teenagers.* Retrieved May 10, 2005, from http://www.useit.com/alertbox/20050131.html

Sullivan, T., Norris, C., Peet, M., & Soloway, E. (2000). *A naturalistic comparison of children's navigation behavior and subjective preferences on two WWW sites.* Paper presented at the Proceedings of the Sixth Conference on Human Factors and the Web, Austin, TX.

Theofanos, M. F., & Redish, J. C. (2003). Guidelines for accessible and usable web sites: Observing users who work with screen-readers. *interactions, 10*(6), 36–51.

Tufte, E. R. (1983). *The visual display of quantitative information.* Cheshire, CT: Graphics Press.

CHAPTER 8

THE DIGITAL GLOBE

Using Google Earth for Virtual Fieldtrips of Coastal Environments

David R. Green
University of Aberdeen, UK

Joanna Mouatt
Marathon Oil UK Ltd.

Virtual fieldtrips have become increasingly popular in educational environments in recent years for a number of reasons. Given declining funds, larger student numbers, and the rising costs of travel, virtual fieldtrips provide a 'common-sense' alternative that can offer students rich resources with which to explore a locality or places further afield. Such experiences reduce or eliminate the costs of traditional fieldtrips, problems of scheduling, weather, and transportation liabilities. Students can also work at their own pace, without the risk associated with being in the field, whilst also having access to and being able to share information. Even where access to the field is still possible, virtual fieldtrips can provide a valuable complementary online resource for a field study area, one which, with the aid of computer technology, offers a student the possibility to visualize and interact with a destination, either prior to or following a visit to the study area.

Digital Geography: Geospatial Technologies in the Social Studies Classroom, pages 147–165
Copyright © 2008 by Information Age Publishing
All rights of reproduction in any form reserved.

One particularly valuable form of virtual fieldtrip involves the study of coastal environments. There are many different commercially available software applications and software utilities to help construct virtual field-trips of coastal environments. These software tools provide an exciting way for students to examine and interact with coastal landscapes using maps and images linked to documents, video clips, and audio. Raising awareness of the importance of GIS and remote sensing in coastal studies is another benefit and forms part of the rationale for a recent proposal to develop an integrated geospatial framework as the basis for the Marine and Coastal Resource Management (MCRM) degree program at the University of Aberdeen in Scotland, UK (Green, 2006). In this chapter, we explore the potential of Digital Globes, such as Google Earth, as the basis for developing a virtual fieldtrip. We use the Ythan Estuary in North East Scotland, UK as an example of how a virtual fieldtrip can be constructed as a teaching, research, and institutional resource for information storage and access.

WHAT IS GOOGLE EARTH?

Google Earth is a form of digital or virtual globe that allows users to search and 'browse' satellite imagery and digital aerial photographs draped over a three dimensional (3D) representation of the Earth. Google Earth employs a simple, yet powerful, set of navigational controls in the form of a dashboard display to permit zooming, panning, rotation, and tilting of an image draped over terrain. Additional functionality includes the possibility to overlay various information themes including roads and settlements onto the background imagery and to create sightseeing tours. The overlay themes are largely aimed towards the provision of more general geographical information for the average Internet user, including streets, businesses, and locations. The imagery for Google Earth is a mosaic of images collected in recent years. Thus, these images are not "real-time." They are obtained from a variety of sources, such as NASA, TerraMetrics and Digital Globe (Google, 2006a). In the UK the high resolution imagery for city areas is provided by Cities Revealed (http://www.citiesrevealed.com/).

Google Earth was first launched in June 2005 (Schutzberg, 2005) as a Beta version (test phase), followed by release of an official and stable version in January 2006 (Version 3). Version 4 of Google Earth was launched in June 2006 with a new interface, new features, and an improved image database (Taylor, 2005–2006). There are three levels of Google Earth: a) a free version of Google Earth, b) Google Earth Plus, which has an annual cost of $20 and provides GPS device support and the facility to import spreadsheets, some drawing tools and improved printing quality, and c) Google Earth Pro, which at the time of writing has an annual fee of $400 and is aimed at the professional and commercial user providing measurement and additional annotation tools, the possibility to overlay many more image types, and enhanced printing.

The popularity of Google Earth has led to the development of a number of support websites (e.g., Google Earth Blog: http://www.gearthblog.com/, Google Earth Hacks: http://www.googleearthhacks.com/downloads) and several related software utilities (or "add-ons") allowing users to add their own thematic layers (e.g., Sketchup, GE Path and GE Tools). This has added considerable power and potential to the initial Google Earth concept. A number of other websites now provide access to information and specialized examples that demonstrate some of the wide ranging potential applications of the Google Earth concept (e.g. Google Ocean: http://www.justmagic.com/GM-GE.html). With the availability of such tools, anyone with a relatively basic knowledge of computing and geographic information can add georeferenced map layers and images of their own to enhance and add value to the layers of information already in Google Earth. In effect this turns Google Earth into a simple Geographical Information System (GIS) that provides users with the capability to develop a customized geographical information system for any area, at any scale, for any part of the world.

The Google Earth Interface

To facilitate ease of use by a wide range of different users, the Google Earth interface is designed to be simple and intuitive. The Google Earth interface provides a basic set of navigational controls to allow the user to zoom and pan around the view, as well as the ability to locate a specific place of interest using the search function. There are three 'panes' to the interface Window; one with the legend, one with the map and image data, and the third containing the navigational tools. A drop down menu-bar resides at the top of the screen. The legend pane also contains a search tool at the top to allow the user to enter a placename in the dialogue box. Upon clicking the *Go* button, the user is "flown" to the location at the highest resolution of the imagery available. The flight may start from the globe view or from any other location or scale where the user has been positioned. Selected places can be marked using a pinpoint option and the location saved for subsequent retrieval in My Places. Additional default themes can also be toggled on and off. (See Figure 8.1)

The 3D terrain feature provides the option to define vertical exaggeration, which enables the user to obtain an enhanced view of the Earth's surface. To generate a 3D view of the terrain, the tilt tool can be adjusted using a slider control (+ or –) until a satisfactory view is seen. To emphasize

Figure 8.1. Google Earth Query: Mylor, UK and Mylor Yacht Harbour, Cornwall, England.

the view of the terrain, and to augment relief features, selection of *Tools > Options > Vertical Exaggeration* provides an opportunity to set the vertical exaggeration to a maximum of 3× (a widely accepted value that neither makes the terrain look too flat nor too mountainous). Access to additional functionality, such as adding images and map layers, is also facilitated through use of the drop-down menus.

There are a large number of themes that can be enabled within the Google Earth interface, such as roads, borders, geographic features and locational information, e.g. restaurants and coffee shops. It is also possible for users to add placemarks and to overlay their own information such as thematic map layers and georeferenced images. Furthermore, if a series of placemarks are located it is then possible to tour these; flying from one point to the next. In effect, Google Earth qualifies as a GIS *mashup* or application that combines information (or content) from more than one source into an integrated user experience.

Add-Ons to Google Earth

The basic Google Earth interface and means of information delivery to the user can be extended using a number of additional utilities that provide the opportunity to add more detailed layers of maps and imagery to a view. Georeferenced maps and imagery can be added relatively easily to provide the basis for a simple local GIS. This is facilitated through a combination of the interface and menus as well as utilities, some of which originate with various Internet support sites. Shape2Earth (Beerman, 2006) is one such application, providing the capability to convert GIS data in the common shapefile format to the Keyhole Markup Language (KML) used by Google Earth. This enables the presentation of GIS data within Google Earth without having to purchase Google Earth Pro. Another example is GE-Graph (Sgrillo, 2006), a Windows-based application that enables the user to plot data within Google Earth as a bar graph. Details of GE-Path and GE-Graph can be found at http://www.sgrillo.net/googleearth/.

The addition of different classes of 3D objects to the surface of a visualization is quite common among many visualization software packages. Google SketchUp, for example, is a free version of SketchUp 5 (http://sketchup.google.com/) which allows for the creation of 3D objects, complete with texture shading and lighting effects. It is therefore possible to design and create a wide range of 3D objects that can be placed within the Google Earth view. Alternatively, users can take advantage of the library of objects already created by other users, such as a building or a wind turbine.

Google Earth Blogs

Support for Google Earth has grown very quickly and there are now a number of Google Earth 'Blogs'. Two of these are the Google Earth Blog at http://www.gearthblog.com/ (Taylor, 2005–2006) and Ogle Earth at http://www.ogleearth.com/ (Geens, 2006). Both provide a wealth of useful information such as tips, news, 3D models, the use of GPS, and sightseeing. They also provide a large number of examples of how Google Earth is being used everyday in a wide variety of ways. Examples cited by the Google Earth Blog (Taylor, 2005–2006) include near real-time severe weather data for the USA, global fire data, coordination of emergency response teams, and the planning of the route for the movement of the Ramses II statue in Egypt in August 2006.

A particularly interesting example of how Google Earth can be used as a web-based GIS is provided by the Timoney Group (2006), who have retrospectively mapped damage to petroleum industry installations within the Gulf of Mexico caused by hurricanes Katrina and Rita in 2005. Google

Earth is used to present information on the oil and gas wells and pipelines within the area and those that were damaged or destroyed by the hurricanes, maps of the hurricane paths, a wind intensity chart, and also bar charts depicting the November 2004 and November 2005 production data statistics. In addition, the Timoney Group (2006) have developed a buffer tool that enables the user to place a buffer around either of the hurricane paths or along the Gulf Coast. Information 'bubbles' for each layer provide layer metadata, while the placemark description for the buffer presents statistics on platforms damaged or destroyed within the buffer zone and the companies affected. The potential of Google Earth as an educational resource is clearly highlighted by Jenkins (2006a), where a number of potential Google Earth-based lessons are provided for download. These include an exercise on the decision-making processes involved in selecting wind farm sites in Devon (http://www.juicygeography.co.uk/gewind.htm).

Google Earth as a GIS

Is Google Earth a GIS? Based on some typical criteria that might be used to define a GIS, (e.g., provision of access to information, thematic layers that can be toggled on/off, tools to control panning and zooming, and 3D visualization), Google Earth can be considered a simple form of GIS because it allows the user to input, store, retrieve, display, and output both maps and images. In addition, Google Earth provides the means to overlay thematic layers, zoom, pan, rotate, make measurements, and employ simple 2D and 3D visualization tools. Albeit a simple example of a GIS, Google Earth nevertheless provides a basic element of GIS functionality for navigation, retrieval and visualization. In addition, this functionality can be extended to provide the means to customize the information for any location in order to create a geographical resource. The primary limitations of Google Earth as a GIS are the lack of capacity for geospatial analysis of large amounts of data and the lack of access and control of the underlying data.

THE POTENTIAL OF GOOGLE EARTH IN AN EDUCATIONAL ENVIRONMENT

Google Earth has the potential to be used in a number of different ways in an educational environment. The simplest of these is to take advantage of the basic Google Earth functionality to "visit" chosen locations and then generate a sightseeing tour using the place markers and the fly-through functionality. For example, if a student enters "Mylor" in the *Fly To* search box he or she will be "flown" to Mylor Yacht Harbour in Cornwall in the Southwest of Great Britain. Additional layers or themes such as roads and settlements can then be added by switching or toggling on/off the radio buttons in the legend pane. The scene can then be zoomed, panned, rotated, and tilted to yield an optimum view of Mylor Yacht Harbour (see Figure 8.1). Using the available tools, functionality, and the mouse to control the view, the user is then able to 'fly' around the locality examining the physical and human geography of the location. By adding place markers, a sightseeing tour of the area can also be defined by highlighting places of interest such as the marina, yacht club, beach area, and car park. Together with contextual information, it is also possible to construct exercises that investigate geographical location such as a marine spatial planning exercise for sighting a new marina, exploring coastal habitat, examining coastal user conflicts, an historical study, or the environmental impact of pollution.

A more elaborate use of Google Earth involves the development of virtual fieldtrips. The virtual fieldtrip can be customized to suit a wide variety of subject matter. The possibility to enhance the basic Google Earth functionality through the addition of multiple layers or themes means that the fieldtrip can also use the concept of GIS as part of the exercise. By using some of the available "add-on" utilities, one can create a customized GIS for a chosen area by adding map and image overlays, links to text and spreadsheet documents, and placing 3D objects in a landscape. In this way, Google Earth can be used to develop a geographical information resource for an area, or a local information system. Once bundled up, such a resource could be freely distributed to other users (assuming copyright permits) and accessed through a common interface without the need for desktop GIS software. The bundle can include multiple layers, 3D objects, links to other documents, and sightseeing tours.

Many examples have already been developed that can be used as information resources, virtual fieldtrips, and teaching exercises (e.g., http://www.digitalgeography.co.uk/archives/2006/02/batsworthy-cross-wind-farm-proposal/; http://www.scapesite.org/GoogleEarthPages/index.htm and http://www.cornwallwildlifetrust.org.uk/reserves/google_earth.htm). Combining a Google Earth bundle with additional tutorial documents provides a teacher in primary, secondary and tertiary education with a very useful and accessible resource that can be used to introduce students to a remote area that it is not possible to visit in the field. Examples might include visits to a geographical location such as a city, mountain range or volcano, a geographical study such as Lewis and Clark's expedition across the USA (http://www.gearthblog.com/blog/archives/2005/11/), or an exploration of the role of geography in determining the best location for something such as a windfarm (http://www.digitalgeography.co.uk/archives/2006/02/batsworthy-cross-wind-farm-proposal). A fly-through can be used as a simple virtual tour of an area with specific locations being highlighted using links to text documents, ground-based photographs, or panoramas. In the context of an aerial photo interpretation or digital image processing exercise such a tour can also be useful for simple ground-truthing. The potential of Google Earth as an information resource, as a teaching tool, a fly through and as a distributed GIS (with many of the inherent elements of a basic definition of a GIS) makes Google Earth a very powerful toolbox indeed.

A VIRTUAL FIELDTRIP RESOURCE FOR COASTAL STUDIES: THE YTHAN ESTUARY EXAMPLE

In this example, we describe how the utilities discussed earlier can be applied to develop a customized virtual fieldtrip application for an estuarine environment. This particular exercise is part of a larger project on the Ythan Estuary in Northeast Scotland, 'YthanView' (http://www.abdn.ac.uk/mrm/YthanView/index), to develop an online Geographical Information System (GIS) to serve as an institutional map and image-based information resource for the research, teaching, and consultancy community at the University of Aberdeen, as well as a virtual fieldtrip to the estuary and the Sands of Forvie, the surrounding sand dune system and a Scottish Natural Heritage (SNH) managed nature reserve. Similar examples could also be developed as exercises and/or projects for school pupils at the pre-university level. Already there are a number of UK websites that provide interesting examples of Google Earth based studies for secondary school students (e.g., http://www.digitalgeography.co.uk/). Currently, the Ythan Estuary project example is being included as one of a number of Google Earth-based GIS examples for coastal management studies as part of a Coastal Zone Youth Club module within the Interreg IIIB COREPOINT project (http://corepoint.ucc.ie/). There is also potential to develop this as a training resource that could be an element of capacity building for Integrated Coastal Zone Management (ICZM). To date the work undertaken has included gathering together of a wide range of geographical data sources, GIS datasets, and a virtual reality (VR) application using VRML and VRML viewers described by Green and King (2002) and Green and Bojar (2005).

Constructing the Ythan Estuary Project within GoogleEarth

The Ythan Estuary project outlined below was created in Google Earth version 4, yet all of the features described also work within version 3. Data used within this project were obtained from the Department of Geography and Environment at the University of Aberdeen, which has been involved in numerous studies of the area. Varieties of data types have been utilized including Ordnance Survey (OS) raster map tiles, digital imagery (aerial and ground-based) and weed mat and sediment maps that were originally created in ESRI's ArcView GIS 3.2 software. The 1:50,000 color raster OS map tile was created from map tiles NJ82 and NK02, which were originally obtained from Edina Digimap (http://www.edina.ac.uk/). The aerial photograph mosaic and weed mat maps were produced by Green and King (2002). Weed mat maps were digitized from aerial photography for each respective year at scales ranging from 1:3000 to 1:7500. Sediment map layers were digitized by Chan (2003). The ground-based digital images (JPEG) were originally obtained during a fieldtrip in January 2005 for ground-truthing of vegetation types within the Sands of Forvie National Nature Reserve (NNR). For convenience, and to provide practical information for the reader, the procedures used are given here as paths rather than describing each step in full. For example, select Add, followed by Placemark opens a New Placemark window and is described simply as {Add/Placemark/New Placemark}. The descriptions for this example are provided as illustrations.

Placemarks

Placemarks allow users to add/highlight points of interest within Google Earth for their personal use or for distribution to the wider Google Earth community. A placemark can be added using the following procedure {Add/Placemark/New Placemark}. The coordinates (degrees, minutes and seconds), description, color, type of marker (the default is the 'push-pin' icon) and altitude of the placemark can be assigned. It is also possible to add images and links to other web pages as part of the placemark description, which requires some knowledge of basic XHTML coding. For example, in the case of the Forvie Dune Grass image the following code <img src = "http://static.flickr.com/59/225170038_bf28f0769e.jpg" / > sources the image from a website that has been designed for sharing digital images (http://www.flickr.com). Alternatively, the images can be sourced directly from the computer hard-disk (e.g. <img src = "C:\Documents and Settings\Desktop\Ythan Maps\Forvie Images\IMG_0009.jpg" / >). It should be noted, however, that sharing of the placemark is only possible if the image is bundled in the KMZ file for the project.

Preparing Image Overlays

Image Overlays provide a means by which the user can enhance the satellite imagery of Google Earth for their own interest or to facilitate use by a wider community. Image file types that can be added directly to the standard free version of Google Earth include the following extensions: .jpg,

.tiff, .png and .gif. Google Earth Pro, however, also allows the addition of many more file types including GeoTiff and Erdas Imagine .img files (Google, 2006b). If the bounding coordinates of the image are known these can be assigned when adding the image, but it is also possible to manually resize, move and rotate the image using the green 'handles' that appear when the properties box of the file is opened. In this project the Ordnance Survey (OS) map tile and the aerial photograph mosaic for the Ythan Estuary were added to Google Earth as image overlays, as will described below.

The OS Digimap file available was in TIFF format, which can be loaded directly into Google Earth as an Image Overlay. To ensure that the raster map is placed in the correct geographical position within Google Earth, the coordinates of the top-left and bottom-right corners (i.e. the north, south, east and western extents of the image) must be known. This information was obtained by opening the file within the Erdas Image 8.7 digital image processing software and consulting the layer properties menu option. An image overlay in Google Earth is added using {Add/Image Overlay/New Image Overlay/'File to Overlay'}. The file is then given a title, description and the location set. To place the image in the correct position within Google Earth the coordinates must be in degrees, minutes and seconds as obtained from Erdas Imagine 8.7, rather than British National Grid (BNG) coordinates. Conversion from grid coordinates to degrees was achieved using the OS Coordinate Transformer for Great Britain (see http://gps.ordnancesurvey.co.uk/convert.asp).

Once all of this information was completed the image was added to Google Earth by clicking OK. For the aerial photographic mosaic, the same procedure was followed, but first it was necessary to convert the original image from an Erdas Imagine file (.img) to TIFF format, since only the free version of Google Earth was utilized for this project. After opening the aerial photographic mosaic in Erdas Imagine 8.7, it was then possible to save the file as a TIFF {Save/Top Layer As}. To reduce the size of the final file loaded into the Google Earth project, the image was then converted into the JPEG format. Also, within Erdas Imagine 8.7, the bounding corner coordinates were noted, and duly converted in the same way as described above for the OS map tile. If the imagery available has no associated coordinate data then JPEG images can be added to Google Earth, as described above, before manual manipulation, using the green 'handles' that define the boundaries of the image when viewing the properties, to ensure correct positioning.

Adding ESRI ArcView Shapefiles to Google Earth

Shape2Earth is a software plug-in developed by Tim Beerman (Beerman, 2006) for use in MapWindow GIS, an open source GIS (http://www.mapwindow.com/) (See Figure 8.2). Shape2Earth can be used to con-

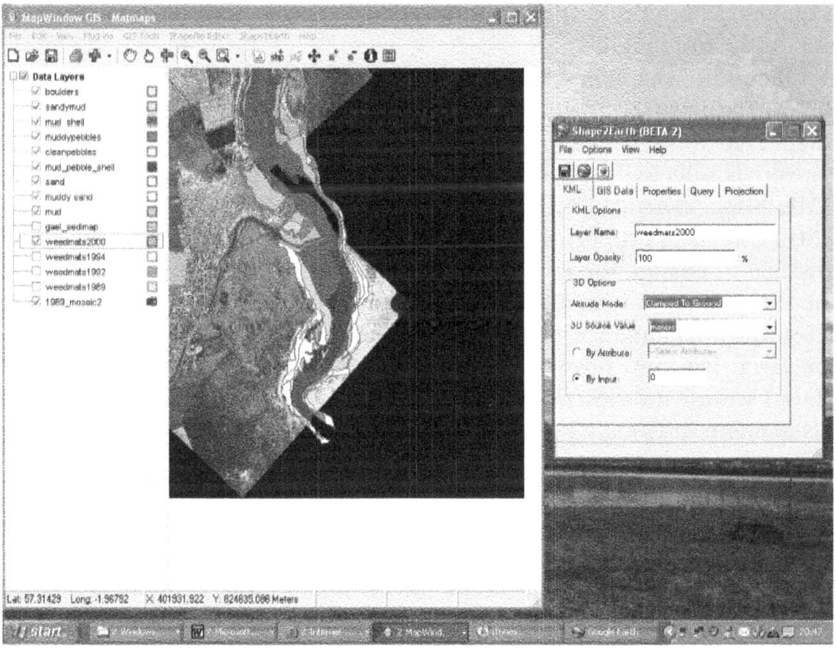

Figure 8.2. Screenshot of the MapWindow GIS and Shape2Earth interfaces.

vert ESRI shapefiles (.shp) to KML, therefore allowing GIS data, which is often in the shapefile format, to be viewed within Google Earth. For this example, several shapefiles relating to the Ythan Estuary were converted into KML format using Shape2Earth, namely the weed mat maps and the sediment maps. First, a shapefile was opened within MapWindow GIS (e.g., 1989 weedmat). Projection information for the shapefile is required, but if no projection was set at the creation of the shapefile it can be assigned within the MapWindow GIS software.

The following procedure can be used to assign a projection to the shapefile {GIS Tools/Vector/Assign Projection to Shapefile/Open File/ 'Required File'/Choose Projection/Category/Group/Name}. In the case of the shapefiles used for this example, the *Category* was set as Projected Coordinate Systems, *Group* was defined as National Grids, and finally the *Name* was set to British National Grid. Reprojection of the file to ensure compatibility with Google Earth, which uses the WGS84 datum projection is also required. This can be achieved through the following procedure {GIS Tools/Vector/Reproject a Shapefile/Open File/'Required File'/Projection to be Applied/Category/Group/Name}. For all shapefiles, the *Category* should be set to Geographic Coordinate Systems, *Group* set to World and the *Name* set to WGS 1984. On clicking OK, the file is saved as filename_Reprojected (e.g. weedmap1989_Reprojected) in the same location as the original file.

Once the shapefiles have been re-projected, it is possible to use Shape2Earth to convert from the shapefile format to KML. Ensuring Shape2Earth is selected in the plug-in menu, and the shapefile you wish to convert is selected in the project legend, select {Shape2Earth/Export to KML/Shape2Earth}. There are five tabs within the Shape2Earth interface: KML, GIS Data, Properties, Query and Projection. In the *KML* section, it is possible to change the layer name, control the layer opacity and also how the layer will appear in 3D in Google Earth. From *GIS Data*, it is possible to select the attribute by which to identify the specific points/polygons within the layer. It is also possible to specify the attributes that will also be displayed within the description for each polygon when selected from the *Places* menu in Google Earth. The *Properties* function can be used to assign the marker type for points within Google Earth, but this feature is disabled for polygons so in this case was not utilized. The *Query* function can be used to load the attribute data for the file of interest, which can then be queried/sorted to give some order to the data (e.g. alphabetical), which allows individual features to be identified much more easily once loaded into Google Earth. *Projection* details the projection information for the shapefile. If there is no projection data shown then this must be assigned, using the procedure detailed above, otherwise conversion from Shapefile to KML will not occur. For conversion, one of three options can be selected from the top left buttons: *Save, Save and Load* or *Load as a network link*. This final option can be useful when work is incomplete on a file as changes can be made and viewed in Google Earth without having to save and reload the file after each change is made. Once the data layers are loaded into Google Earth, it is also possible to alter properties such as description, opacity, color and altitude by right-clicking and selecting *Properties* from within the *Places* pane.

Adding Graphical Data to Google Earth

Another freeware program developed to aid the visualization of data within Google Earth is GE-Graph, created by Sgrillo (2006) (See Figure 8.3). GE-Graph is a Windows-based application that can be used to generate geospatial bar graphs either from KML files, saved by Google Earth, or from data entered directly into the application, which can then be exported into Google Earth (Sgrillo, 2006). For this example, the data for each sample were entered manually into the GE-graph application (i.e. latitude, longitude, sample name, and value). The latitude and longitude were required to be in decimal degrees for input into GE-Graph, therefore an online calculator was used to convert the location of each sample from

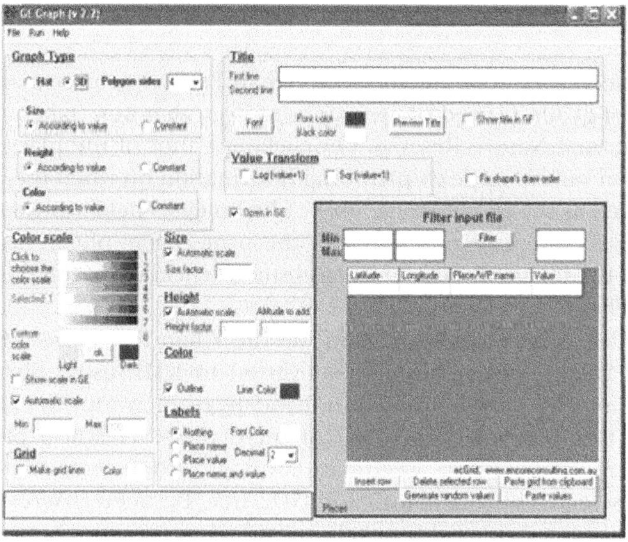

Figure 8.3. Screenshot of the GE-Graph interface.

degrees and minutes to decimal degrees (see Johnston, 2005). The physical bars of the graph can be produced in a variety of shapes, from circles to squares, or as polygons. The size or height of the shapes can be set to correspond to the value of the placemark. Assignment of a label (e.g. place name and/or value) for each placemark and production of a title and color scale as screen overlays are just some of the other features available within this application.

Ythan Estuary Project

The Ythan Estuary Project files were bundled together as a zipped file (kmz) that can be loaded into Google Earth. Once a user loads the file into Google Earth and double clicks on the Ythan Estuary Project link, he or she will 'fly' to the default view of the project files. An information bubble appears to provide some information about the project and links are provided to web pages that contain relevant online information sources such as the M.Sc. dissertation title pages and abstracts of Chan (2003) and Dunne (2003). The user also has access to the OS map tile and a 1989 Aerial Photo Mosaic files; indicating how image overlays appear within Google Earth. Both are of particular interest here given that the imagery for this particular study region is unfortunately of poor quality, not only with respect to image resolution, but as it is also an area severely affected by cloud cover.

Two placemarks were added to the Ythan Estuary project that show the locations of ground-based photographs of flora that were taken within the Forvie National Nature Reserve. On clicking either of the placemarks, an information bubble appears providing information on when the image was taken as well as the actual image itself. The image is held at www.flickr.com to achieve relative ease of sharing of the project file, plus it also enables viewing of the image directly at the website by clicking on the link "See the image in Flickr."

Shapefiles detailing weed mats within the Ythan Estuary for 1989, 1992, 1994 and 2000 were individually converted to KML using Shape2Earth. The files can therefore be viewed individually or in groups to examine the extent of the weed mats for each of these years. This also provides an opportunity to compare weedmat coverage from year to year. The attributes selected during conversion (ID, Area, Perimeter, Acres and Hectares) can also be displayed for each individual weedmat polygon by selecting the ID number from the *Places* menu. A *Sediment Maps* folder contains the individual converted shapefiles for each of the observed sedimentary types within the Ythan Estuary. By toggling the folder on, all the sediment types are displayed. Each individual polygon of a sedimentary type is accompanied by an attribute table that details the ID, Type, Area, Perimeter, Acres and Hectares. Again the attribute table is displayed within the placemark description obtained by clicking on the polygon ID.

Finally, a short tour was created where the user flies from the mouth of the Ythan Estuary to its furthest reaches as displayed within the aerial photo mosaic. The Ythan Tour enables the user to obtain an impression of the Ythan Estuary without setting out into the field (See Figure 8.4). This was achieved by placing eight placemarks (with no icon) along the river. The variations in heading and tilt of the view were set before adding each placemark so that it was automatically saved within the placemark KML code. Prior to starting a tour, it is best to optimize the tour settings by establishing an appropriate tour speed and tour pause settings. This is accomplished by the following procedure {Tools/Options/Control/Speed or Tour Pause}. These settings are of course subjective and should be adjusted according to need. A user should also ensure that *Terrain* is selected within the *Layer* panel (below the *Places* panel) and that in the *View* tab of Google Earth *Options* the *Vertical Exaggeration* is set to 3, therefore providing a 3D representation of the Ythan Estuary. To play the tour, the Ythan Tour folder is selected within the *Places* pane (it should turn grey) and the user clicks the play arrow situated at the base of the *Places* pane. The result is that the Google Earth tour represents a very rapidly produced alternative to a Digi-

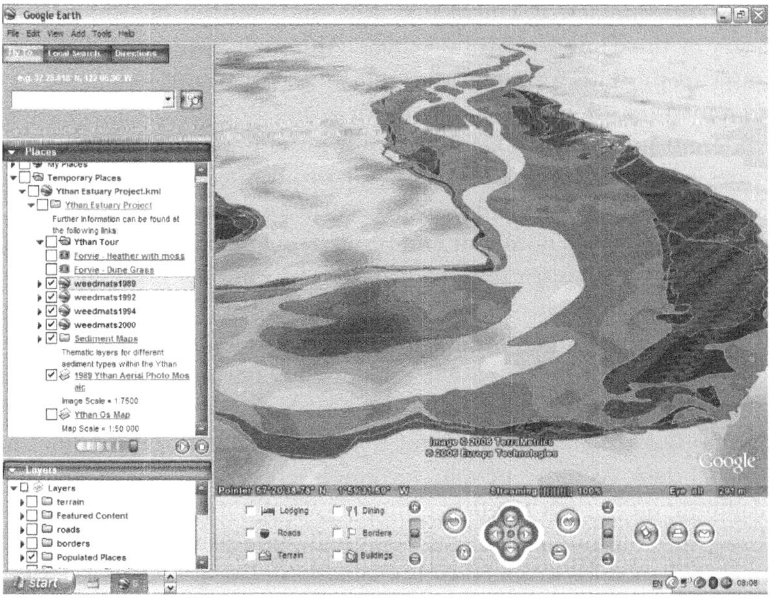

Figure 8.4. Portion of the "Ythan Tour" showing the terrain, sediment layers and several weedmat layers.

tal Elevation Model (DEM) for the area, afforded by the terrain feature of Google Earth that allows the user to view the 3D landscape of an area.

DISCUSSION

The Ythan Estuary project files created demonstrate that visualization of geospatial data can be successfully achieved using Google Earth. Furthermore, the ability to bundle project files into a zipped file format (kmz) facilitates file sharing and distribution of customized application files. The addition of the OS map tile and aerial photograph mosaic as overlays in Google Earth also provides greater resolution data for the area, which in this example location unfortunately suffers from relatively poor base imagery. This provides the opportunity to create an improved contextual setting for visualization of the sediment and weed mat map data for the area. Furthermore, the ability to display the attributes relating to specific shapefiles within Google Earth allows the user not only to visualize the data presented, but also to retrieve a variety of additional contextual information. It is also envisaged that this feature could enable the user to perform simple GIS-like queries on the data, such as that demonstrated in the assessment of the impact of Hurricanes Rita and Katrina on the petroleum industry by the Timoney Group (2006). Although not undertaken for this example, it presents a significant opportunity for future investigation and development of the use of Google Earth as an Internet-based GIS interface for different study areas. The Tour facility present within Google Earth also provides a relatively quick and easy way to produce a 'fly-through' of an area. It is anticipated that this feature may be of interest to a wide range of users with numerous potential end-results, such as virtual fieldtrips, field-trip reconnaissance prior to a fieldtrip, or even simply detailing places visited during fieldtrip.

It is clear that the additional resources provided by Shape2Earth and GE-Graph significantly increase the potential of Google Earth as a resource for the presentation of geospatial data. It is also evident that 'digital globe' applications provide useful tools for the visualization and dissemination of geospatial data. As more people become interested in the geospatial data available through such media, further improvements, developments, and add-ons will continue to appear; potentially increasing the ability to perform spatial analysis. As noted by Butler (2006), Google Earth has revealed the potential of 'digital geography' to the masses, which is evident from the range of content produced by everyday users of the software and reported on Internet sites such as Google Earth Blog (Taylor, 2005–2006) and Ogle Earth (Geens, 2006). For example, during the final stages of this project research, an application to produce KML for spreadsheet data has been described by Jenkins (2006b), providing a timely reminder of the constantly evolving nature of the technology. Other examples of the range of uses to which Google Earth can be applied include tools to generate panoramic images (http://digitalurban.blogspot.com/2006/09/how-to-create-aerial-panorama-from.html), as well as animations to monitor spatial change over a period of time.

SUMMARY AND CONCLUSIONS

Geospatial data is relevant to everyone and web-based digital globes, such as Google Earth, have effectively opened up the world of geographical information to more users by providing access to data that was previously available only to a minority of people. The dissemination and visualization of geospatial information is greatly facilitated by the use of the Internet and freely available web-based applications. Furthermore, the speed at which these 'digital globe' applications are evolving, through the production of new themes, content and integrated tools, testifies to their popularity as a medium through which people can explore or map their own personal world.

This chapter explored the potential of Google Earth for the visualization and dissemination of geospatial information in educational environments. It has been shown that a wide variety of geospatial data sources, such as georeferenced digital photographs, digital maps, points, shapefile and associated attribute data can be readily integrated with Google Earth with exciting results. The use of a user-friendly application like Google Earth for the presentation of geospatial data has widespread implications for many audiences. This is of particular importance in educational and public participation exercises where the emphasis should be on data visualization and interpretation. Furthermore, the user-friendly and visually striking interface provides a captivating medium through which to view geospatial data. Whilst there are a number of similar applications now available (e.g., NASA's World Wind: http://worldwind.arc.nasa.gov/ and Microsoft's Virtual Earth: http://www.microsoft.com/virtualearth/default.mspx), the lack of such user-friendly interfaces and comparable functionality make Google Earth the most appealing and useful toolbox. The "add-on" tools available to enhance the basic information available within Google Earth are not yet as user-friendly as they could be, but their presence as well as the support offered by current 'Blog' sites suggests increasing opportunities to expand the basic functions of Google Earth for local GIS purposes.

This chapter has provided a brief excursion into the possibilities afforded by Google Earth and related applications for coastal field studies. Giving students the potential to access coastal environments with the aid of computer technology adds a very valuable dimension to their school, academic and vocational studies. Depending on the availability of datasets and information for a location, a fieldtrip can take the form of a simple diary, an illustrated walk, an online visual tour, or a very rich and dynamic visual and interactive experience. This provides students with the opportunity to work with data, to become actively involved in practical field exercises, and even to participate in the learning process by designing their own fieldwork investigations. Our future work will concentrate on developing the YthanView project into a more comprehensive resource, using more data sources, whilst also examining the potential for development of facilities to enable querying and animation of datasets. It is evident that digital globe applications, such as Google Earth, represent a significant resource in the presentation and distribution of geospatial information that has proven to be instantly popular with the general Internet user. The Google Earth phenomenon represents a fusion of GIS, the Internet and the general population with considerable potential in an educational context for all levels.

REFERENCES

Beerman, T. (2006). *Interactive Earth (28/08/06), Colorado Springs, Colorado, USA.* Retrieved April 10, 2007, from http://www.interactiveearth.blogspot.com

Butler, D. (2006). The web-wide world. *Nature, 439,* 776–778.

Chan, K. H. (2003). *Sediment and Macro-algae relationship of the Ythan Estuary: A remote sensing & GIS approach.* Unpublished master's thesis, University of Aberdeen, Scotland, UK.

Dunne, B. (2003). *The relationship between algal mats and wading bird populations in the Ythan estuary, North East, Scotland.* Unpublished master's thesis, University of Aberdeen, Scotland, UK.

Geens, S. (2006) *Ogle Earth: A blog about Google Earth.* Retrieved June 12, 2007, from http://www.ogleearth.com/

Google. (2006a). *Google Earth.* Retrieved April 11, 2007, from http://earth.google.com/ earth.html

Google. (2006b). *How can I import GIS imagery into Google Earth.* Retrieved April 11, 2007, from http://earth.google.com/support/bin/answer.py?answer=21332 &query=projection&topic=&type=f

Green, D. R. (2006). Promoting the development of a geospatial technology framework for a marine and coastal resource management degree programme: an example from the University of Aberdeen, Scotland, UK. In C. D. Woodroffe, E. Bruce, M. Puotinen, and R. A. Furness (Eds.) *Proceedings of CoastGIS 2006, The 7th International Symposium on GIS and Computer Cartography for Coastal Zone Management,* 25–36.

Green, D. R., & Bojar, K. (2005, July). YthanView–Visualizing an estuary and virtual fieldwork at the Ythan Estuary, Scotland, UK. In D. R.Green (Ed.) *Proceedings of CoastGIS 2005, The 6th International Symposium on GIS and Computer Cartography for Coastal Zone Management,* 148–161.

Green, D. R. and King, S. D. (2002). Ythan Estuary-The distribution of weedmats in the intertidal zone, University of Aberdeen, Scotland, UK

Jenkins, N. (2006a). Google Earth in the classroom. *Juicy Geography.* Retrieved April 10, 2007, from http://www.juicygeography.co.uk/googleearth.htm

Jenkins, N. (2006b). Online spreadsheets and KML. The possibilities are endless! *Digital Geography.* Retrieved April 10, 2007, from http://www.digitalgeography.co.uk/archives/2006/08/online-spreadsheets-and-kml-the-possibilities-are-endless/

Johnston, E. (2005). Degrees, minutes, seconds to decimal degrees calculator. *Satellite Signals Ltd.* Retrieved April 10, 2007, from http://www.satsig.net/degrees-minutes-seconds-calculator.htm

Schutzberg, A. (2005). Google Earthrise. *Directions Magazine.* Retrieved April 10, 2007, from http://www.directionsmag.com/article.php?article_id=898&trv=1

Sgrillo, R. (2006). G*E-Graph: Graph for Google Earth (26/08/06), Cocoa Research Center CEPLAC/CEPEC), Ilheus, Bahia, Brasil.* Retrieved April 10, 2007, from http://www.sgrillo.net/googleearth/index.htm#GEBarsGraph

Taylor, F. (2005–2006). *Google Earth Blog.* Retrieved April 10, 2007, from http://www.gearthblog.com

Timoney Group. (2006). *The energy impact of Rita and Katrina, The Timoney Group, Denver, USA.* Retrieved April 10, 2007, from http://www.gulfimpact.com/

PART III

EXPLORATIONS OF THEORY AND RESEARCH ON GEOSPATIAL TECHNOLOGIES IN EDUCATION

CHAPTER 9

DIFFUSION OF INNOVATIONS THEORY

Framing IT and GIS Adoption

Shannon White
University of South Florida

Adoption of any new idea or technology happens over time. Everett Rogers' oft-cited (2003) work in Diffusion of Innovations (DI) theory may be used to better understand the processes and persons involved in instructional technology and GIS diffusion in K–12 schools. Rogers (2003) theorizes, "the main elements in the diffusion of new ideas are: a) an innovation b) that is communicated through certain channels c) over time d) among the members of a social system" (p. 36). In this chapter, I outline specific elements of Rogers' theory relevant to the issues of IT and GIS in K–12 settings. Rogers' "categories of adopters" model, discussion of decision-makers, and stages of the innovation process are each significant to the discussion of individual educators regarding critical decisions in the adoption of innovations. In the final section of the chapter, I present results from my research on students' and educators' conceptualizations of GIS in schools (White, 2005).

Digital Geography: Geospatial Technologies in the Social Studies Classroom, pages 169–195
Copyright © 2008 by Information Age Publishing
All rights of reproduction in any form reserved.

DIFFUSION OF INNOVATION (DI) THEORY

According to Rogers (2003), the process of diffusion differs for each innovation. As shown in Figure 9.1, this process is dependent on time and the rate of the process can be very swift or extend over much longer periods. The s-shaped curve for three innovations shown in Figure 9.1 is comparable, but the rate is different for each.

The s-curve for Innovation I in the figure could be likened to Internet adoption in K–12 schools in the U.S. According to a National Center for Educational Statistics (NCES) report, 35% of schools had Internet access in 1994. Eight years later, in 2002, the figure was reported as 99% (NCES, 2004). While that number may represent one computer in the main office of the school, the adoption of Internet access was swift. The percent of computers designated for instructional use with Internet access follows a similar trend. In 1995, the percentage of instructional computers with web access was 8% and by 2002 it climbed to 90% (NCES, 2004). This growth is approximately a factor of nine over that eight-year period. This is explosive adoption and implementation of this innovation.

The s-curve for Innovation II is stretched out over more time. The curve is less "steep" than that of Innovation I. This curve might represent instructional technology adoption. NCES also reported 5,621,000 computers in schools designated for instructional use in 1995. By 2002, that number had grown to 10,711,000 (NCES, 2004). While I do not know how often, when or where these computers were used in the schools these statistics do give

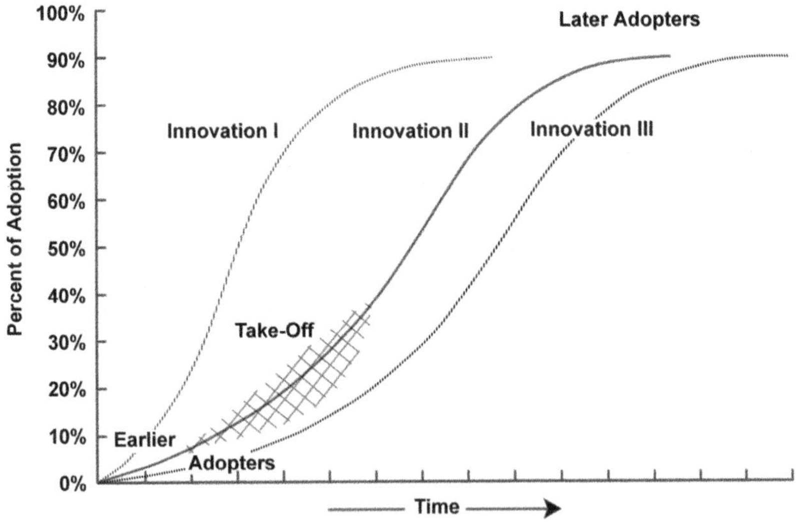

Figure 9.1. The Diffusion Process (After Rogers, 2003).

us a snapshot of the rate of change of instructional technology from a very broad perspective. This growth did double in the seven year period reported, however, it was not as steep a growth rate as was Internet access.

Innovation III shows the slowest rate of adoption and diffusion over time. I suggest that this curve is more congruent with the adoption and diffusion of GIS in K–12 education in the U.S. While NCES does not report on statistics regarding GIS use in schools, there is early research that indicates the number of users and the slow diffusion rate of GIS. Kerski's (2000, 2003) survey of 1520 secondary school teachers represents a sample of the 1800 individual educators in the U.S. that were known to own a GIS software package. In addition, he found less than 5% of U.S. high schools owned a GIS package (Kerski, 2003). This was the first study to tabulate the users of GIS in the education realm. Kerski (2003) compares these numbers to the 500,000 total GIS users worldwide reported by ESRI in 1999. This study has given researchers a starting point to begin to discuss the slow adoption rate of GIS in education compared to the rate in other industries.

Social Processes of Adoption of Innovations

One of the better-known features of Rogers' theory is his model of the categories of adopters of innovations (see Figure 9.2). Within a discussion of innovation theory and instructional technology, Surry and Ely (1997) observes: "Technology's advance may be inevitable, but it is gradual. Instructional technologists should, therefore, look to the potential adopters to show us ways to gradually introduce our innovations into their societies" (p. 41). Understanding the role and perceptions of instructional technologists and

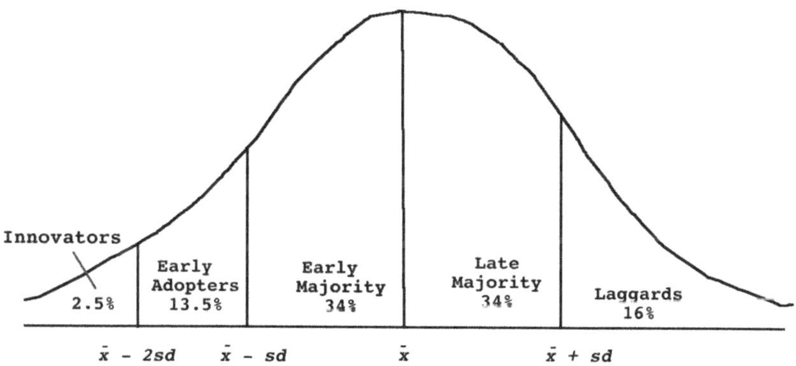

Figure 9.2. Adopter categorization on the basis of innovativeness (After Rogers, 2003).

other potential adopters in educational settings may facilitate improved decision making processes. Rogers labels individuals within the innovation model with five adopter categories: *innovators, early adopters, early majority, late majority,* and *laggards.* These categories are based on their openness to innovation.

Considering the categories of adopters is important to an examination of instructional technology and GIS as most educational institutions have persons who fall into each of these categories. GIS in education is in the earliest stages of this s-curve of adoption and the users at this time are most likely considered *early adopters.*

Overall, instructional technology (IT) diffusion among individual educators might be seen to fall into the latter stages of *early majority* and into the initial stages of the *late majority* classification, where most of the population would have adopted the innovation. There are still individuals who have not adopted instructional technologies for classroom use, and IT is present in schools at varying degrees of a broad range of technologies from educators using TV/VCRs to computers.

Organizational Adoption and Diffusion of Innovations

The types of decisions made in an educational institution are important, but also key to this discussion is the process an organization considers for the diffusion of an innovation. Rogers (2003) explains there must be a "perceived need" of the innovation and a match to the goals of that organization (see Figure 9.3). Once the needs and matched goals are understood then a decision for or against adoption will be made. Next, the organization must carry out that decision. Rogers calls this activity the implementation stage, which would involve a modification of existing structures in the

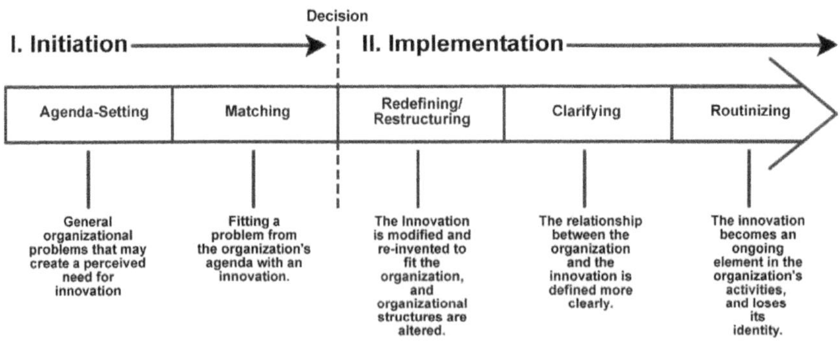

Figure 9.3. Five Stages in the Innovation Process in Organizations (After Rogers, 2003).

organization, clearly defining the relationship of the innovation (IT and/or GIS in this case) and then making the innovation a common practice of the organization (Rogers, 2003).

With regard to instructional technology, most schools have met the initiation stages of the diffusion model. National agendas regarding technology integration in the curriculum can be found in the ISTE National Educational Technology Standards for teachers and students (2000) and No Child Left Behind Act (2001). As educational institutions have moved through the initiation stages they find themselves in the *redefining and restructuring* and *clarifying* stages alluded to by Rogers.

Next, an organization moves into what Rogers describes as matching. "At this second stage in the innovation process, conceptual matching of the problem with the innovation occurs in order to establish how well they fit" (Rogers, 2003, p. 423). The discourse surrounding GIS and educational resources has been inundated with anecdotal accounts of teacher and student use of GIS. This discourse has shown the desire for the adoption of GIS as a tool for educators. If we examine GIS as an innovation within Rogers' (2003) definitional theory of diffusion of innovations, GIS in education is still in the early stages of diffusion. "The relative advantage of an innovation [in this case, GIS in the curriculum], as perceived by members of a social system [K–12 education], is positively related to its rate of adoption" (Rogers, 2003, p. 265).

IMPLICATIONS OF DI THEORY FOR GIS IN K–12 EDUCATION

Examining GIS in education from DI theory suggests that the earliest users of the innovation, GIS, are beginning to spread their knowledge and understanding of it through their social networks. Thus the anecdotal work of these "*innovators* and *early adopters*" functions as a collective voice against the resistance of schools to move forward with the innovation. Kerski's (2003) national study of GIS implementation demonstrates that the *early adopting* GIS "teachers (88%; $n = 342$) believed that the use of GIS makes a significant contribution to learning throughout secondary education" (p. 131). The educators in White's (2005) study and throughout anecdotal and qualitative literature have begun to demonstrate where there is a place for GIS as an innovation in schools. They have revealed and demonstrated matches to curricular objectives and pedagogical concerns in multiple disciplines.

GIS software companies develop new products for their business customers with the need for more power, memory and speed. Industry leader, Environmental Systems Research Institute (ESRI) notes in their product brochure, GIS for K–12 Education: Solutions for Students and teachers that "GIS is a resource-intensive technology. It works best on fast processors with significant RAM and sufficient storage space available" (ESRI, 2004). Schools are left behind in the wake of the new software rollouts because of the failing hardware and inability to maintain the same turnover as the business sector. ESRI in the spring of 2001 released a more powerful and user-friendly version of their software, ArcGIS 8.1 (ESRI, 2004). The hardware requirements at that time were impossible for most public schools to meet. One year later ESRI press released *Mapping Our World: GIS lessons for educators* (Malone, Palmer & Voigt, 2002), a GIS resource book which included a one year license for schools of their ArcView 3.3 software. This software is the predecessor to ArcGIS. While the ArcView 3.3 product is currently still supported, the company will be phasing the 3.3 version out. Subsequently, the second edition of *Mapping Our World* (Malone, Palmer & Voigt, 2005) was released in ArcGIS 9, but only newer school computer labs can support it.

The GIS software on the market today is predominantly Microsoft Windows based. For schools with a majority of Apple computers, which run Macintosh operating systems, the possibility of GIS integration is limited. ESRI's last product that supported an Apple Macintosh computer was ArcView 3.0, yet it is no longer sold and is supported only on the OS 8 and 9 platforms. ESRI in 2004 recognized this issue with the release of a freely downloadable product, known as Arc Explorer—Java Edition for Education or AEJEE.

If the diffusion process model, as seen in Figure 9.1, were to be overlaid onto the organizational decision making process model, seen in Figure 9.3, the stages an organization must move through in order to adopt an innovation would be more clearly seen as dependent on the individual categories of adopters and the rate of adoption (see Figure 9.4). This is relevant to the discussion of software as we move into the decision making stage for an organization and then into the implementation of an innovation if it is adopted.

As GIS is in the initiation stages with respect to education, *innovators* and *early adopters* of GIS in education have worked through the matching processes and restructuring stages for themselves and are helping to clarify the place of GIS in education. At this time, however, GIS in education can claim no more than a 20% adoption rate; and is probably closer to a 5–10% adoption rate. I base this estimate on the correlating stages of decision-making (*agenda-setting* and *matching*) based on the Innovation s-curves. GIS adoption in education, as previously stated, is more closely aligned with the Innovation Curve I and would therefore be at a 5% adoption rate.

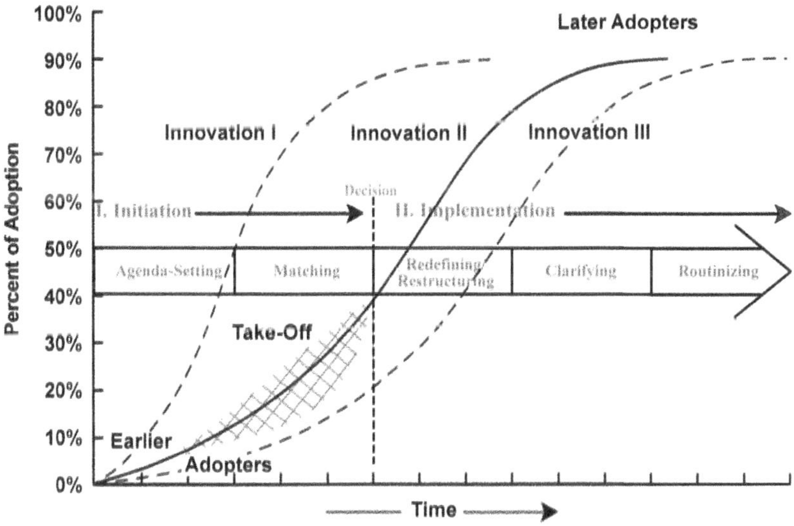

Figure 9.4. Rogers' Rate of Adoption overlaid on the Stages of Organizational Decision Making.

An understanding of how these innovators and early adopters conceptualize GIS globally and GIS in schools may aid in understanding how GIS may or may not benefit existing K–12 schooling and curriculum. Any parallels between the two may illuminate how GIS may be tied to real world experiences. This is particularly important to the research questions of the study, which focused on how educators and students conceptualize GIS in school and 'real world' settings. The findings of this research may be able to inform decision-making in the adoption process.

Interestingly, the software developers and those interested in marketing GIS in education have also begun to shift their products toward Rogers' stages of restructuring based on feedback from the innovators and early GIS K–12 adopters. The software and hardware issues that have been complicated in the matching process are being restructured by companies like ESRI to help organizations move toward the decision to adopt, and in effect accelerate the rate of adoption. The multiple platform solution and increasing number of Internet-based mapping services are enabling GIS to be used in schools where infrastructure and economic concerns are at the forefront in the discussions of adoption of innovative technology. As in Figure 9.4, the shift by the software manufactures toward Internet-based GIS (IGIS) may make changes to the rate of adoption both in time and in number of adopters. This marketing decision could accelerate the slow s-curve of GIS through the use of the Internet, which had a rapid rate of adoption thus more closely matching the GIS s-curve more with the s-curve of instructional technology.

Baker (2005) addressed this issue discussing specific advantages of the use of Internet or web based mapping. When constraints such as hardware and infrastructure are lessened by the use of the Internet rather than computer based mapping, Baker suggested, "the technical preparation time that is now required of teachers [with the use of desktop GIS] could be largely shifted back to focusing on instruction [with IGIS]" (p. 47). This is not to suggest that IGIS is the only solution or even *the* solution to the initiation decisions surrounding GIS. Baker (2005) posited, "The Internet-based environment is ideal for instigating geographic analysis in classrooms where such activities would have otherwise not occurred at all" (p. 47). It is one approach that potentially may aid in the diffusion of GIS as an innovative educational resource.

As others in this volume have noted, there are limitations that must be considered in Internet-based mapping. Internet-based mapping generally does not have the robust analytical capabilities of a desktop GIS. Student exploration of the data they have collected is also limited by web-based mapping. The *innovators* and *early adopters* may find the limitations of a web-based GIS frustrating. Another limitation is that before students can use web mapping it has to be constructed by someone ahead of time (unless they are using existing web based services). These limitations must be considered by educators, web mapping service providers, and GIS software manufacturers.

INTEGRATING THE THEORETICAL FRAMEWORK OF DI INTO RESEARCH ON GIS IN EDUCATION

The following is a presentation of a component of my dissertation research (White, 2005), which integrated the theoretical frames of DI theory, Dual Coding Theory, generated by Paivio (1986) and phenomenographic methodology after Novak and Gowin (1984), and Mavers, Somekh, and Restorick, (2002).

Overview of the Methodology

The study was focused on the experiences, understandings, and ways of thinking about GIS, instructional technology (IT) and the relationship of GIS and IT to schools. This was a multiple case, or multi-case, study of six individuals who had demonstrated equivalent skills in GIS and were either educators or students in K–12 education. The research questions explored the issues of "how" these individuals conceptualize the phenomenon of GIS in schools.

Through comparative analysis of the data (representational drawings, interview and observation) the understandings of these individuals may aid in the decision-making process of GIS integration in K–12 education as well as provide insights for GIS educators and instructional technologists in implementing this tool throughout the curriculum. The methodology of this multi-case study draws on spatial cognition theories, prior research in the fields of geography and education and methods emerging from phenomenographical research.

The qualitative research methodologies of case study, interview, and phenomenography were selected based on the nature of the phenomena investigated. In order to understand how individuals conceptualize and make meaning about a given topic, a qualitative approach is useful. Bogdan and Biklen (1998) specify five features that are key in the decision to use qualitative rather than quantitative analysis. These features include a concern for: a) naturalistic settings, b) descriptive data, c) process rather than outcomes or products, d) inductive analysis, and e) understanding meaning through participant perspectives (Bogdan & Biklen, 1998). These features are important to understanding how the lived experiences, knowledge acquisition, and perceptions of educators and students influence how they understand and think about the larger topic of GIS in schools.

The decision to use multiple case, or multi-case, study research rather than a single examination of one individual's perceptions of GIS, IT, and GIS in schools is based on the assumption that each individual brings a unique perspective to the table that may help to paint a more detailed picture of the research question. Each of the six participants in the investigation represents a separate case.

Paivio's (1991) dual coding theory provides a framework for the research methodology. Paivio's cognitive theory explains how individuals internally code experiences, concepts and phenomena. Paivio identifies the dual encoding modes of both linguistic and imagistic systems that refer to one another. The linguistic elements he calls *logogens* and the imagistic, *imagens*. The connections and referencing between *imagens* and *logogens* are individually constructed, representing their own conceptualizations. This theory is complemented by the concept map research conducted by Novak and Gowin (1984).

The inclusion of representational drawings as data also draws on prior research such as Kevin Lynch's *Image of the City* (1960), an examination of how a phenomenon (a city) is perceived by the subjects of the study (in his case the inhabitants) and represented through drawings or sketch maps. Additionally, Lynch's research was based on spatial perceptions and geographic locations, which have relevance to the study of GIS. This early research, using subject-drawn representations, helped develop an understanding of the different ways in which individuals perceive, experience, and think about a common location or phenomenon. This idea of externalized representations is not limited to geographic studies as it has been the focus of research for educational psychologists and instructional technology researchers as well. In Vosniadou and Brewer's multicultural study, *Mental Models of the Earth* (1992), the combined techniques of representation and interview yielded statistically significant and insightful views into American and Greek childrens' developmental mental representations as constructions and changing conceptions of the earth.

Novak and Gowin (1984) provided the foundational research in concept mapping as a method to understand individual's cognitive maps. Novak and Gowin (1984) found "concept mapping is a technique for externalizing concepts and propositions (p. 17). They began by evaluating concept maps quantitatively based on Ausubel's (1963) cognitive learning theory. Novak and Gowin offer one of the scoring criteria for concepts maps they used based on the following categories: propositions, hierarchy, cross links and examples (1984, p. 36). This scoring technique however is intended for educators to evaluate student work in the classroom. Subsequent research in the United Kingdom by, Somekh and Restorick (2002) has built upon the models of Novak and Gowin (1984).

Mavers, Somekh, and Restorick's (2002, 2003) research on school children's conceptualizations of Information and Communications Technologies (ICT) provides a foundation for methods used in this study. The researchers used concept maps or "mind maps" as the starting point of the research. The student-produced representations were then used as the basis of the interviews conducted for deeper analysis. From this combination of data sources, categories of understanding of ICT emerged. In the British research, the two categories that surfaced were "Spheres of Thinking (SoT) and Zones of Use (ZoU)" (Mavers et al., 2002; Somekh & Mavers, 2003). These researchers used a combination of quantitative and qualitative methods to identify the patterns in their data. This investigation, a multi-case study focusing only on six individuals, based on the methodological structure provided by Somekh and Mavers and is applied to the conceptual understandings of educators and students with respect to GIS in schools.

Phenomenography

In order to analyze the conceptualizations of the participants, this multicase study is informed by the theory and methods of phenomenography. "The object of [phenomenographic] research is the qualitatively different ways in which people are aware of the world, experience, understand, apprehend and make sense of various phenomena and situations around us"(Marton & Fai, 1999, p. 10). At the core of phenomenographic methods is the interview that may be supplemented by other data such as representations.

Phenomenography originated in Sweden in the 1970s. Original writings in this field began with Ference Marton in 1981. This qualitative approach is empirically based. Phenomenographic researchers seek to categorize the understandings of individuals in order to better understand the phenomena from commonalities found by 'pooling' a set of individual understandings. These categories are often referred to as "categories of description." Marton and Booth (1986) note, "It is a goal of phenomenography to discover the structural framework within which various categories of understanding exist. Such structures (a complex of categories of categories and description) should prove useful in understanding other people's understandings" (p. 34). In phenomenographic literature, this emergence of the categories of understanding (or description) from the data is a key characteristic of the analysis.

> All of the material that has been collected forms a pool of meaning. It contains all that the researcher can hope to find, and the researcher's task is simply to find it. This is achieved by applying the principle of focusing on one aspect of the object and seeking its dimension of variation while holding other aspects frozen. The pool contains two sorts of material: that pertaining to individuals and that pertaining to the collective. (Marton & Booth, 1997)

Beyond the categorization of the ways of knowing, the relationships between the categories of description are also important to the researcher. Phenomenographic data is most often collected through semi-structured interviews but also may be collected from numerous other sources such as drawings, observed behaviors and products of an individual's work (Dall'Alba, 1996; Marton, 1986).

Participants

The study participants were geographically located in the central piedmont region of North Carolina. The geographic constraints were based on the researcher's available time and travel. Students and educators were also selected based on three additional criteria: a) prior GIS experience, b) representation of educators from multiple educational settings and curriculum areas, and c) availability and willingness to participate in the study.

Criteria for Selection

Prior GIS experience was determined by separate means for the students and the educators. The adult candidates for participation were determined by their previous enrollment in a GIS course with a K–12 educational focus at "Atlantic Coast University" (ACU—a pseudonym). The student candidates for participation successfully completed two GIS courses in a local middle school. All of the potential participants had similar GIS training and background. The *GIS in Education* course at ACU was based in constructivist principles. The educators enrolled in this course were taught the GIS technology as well as how to implement it for classroom use. While interdisciplinary application was stressed throughout the course, educators were encouraged to create projects that fit the needs of their own classrooms.

The middle school students had completed at least two 18-week (semester) courses in GIS. The GIS skills acquired by these students were equivalent to those taught in the single semester college course for educators. In addition, the students in the study were taught at least one course in GIS by a teacher who had been enrolled in the first *GIS in Education* course. The teacher, Mrs. King, also became a part of the instructional team of the graduate-level course after completing it. The similarity in pedagogical philosophy and approaches used in both levels provided some consistency while attempting to understand the similarities and differences in the experiences of the participants at both levels.

Requests for participation were sent in the form of a letter of introduction to the study. The initial introduction was followed up by email and telephone requests. Interviews were scheduled at the convenience of the participants (and parents of the students) and the researcher. Each participant was asked to participate in three semi-structured interviews during which the drawing tasks were an initial task.

The final group of educators included two females and two males. Two of the educators (one of each gender) worked in private schools; two (one of each gender) in public schools. Of each pair, one was a classroom teacher, and the other had other responsibilities, among them, instructional technologies.

How Do Students and Educators Conceptualize Geographic Information Systems in Schools?

What follows are results from one portion of the 2005 research. Participant educators, Kathleen, Charles, Natalie and Ben, and middle school students Ernest and CJ (pseudonyms) had each completed two prior interviews and representations prior to this final interview and representational drawing. The findings from this particular question are reported since they are most closely connected to curriculum and social studies. The protocol for the task was as follows:

Protocol for Interviews:

Today I would like you to think about GIS in Schools.

This should be the title of your map. [Pause]

Where is GIS in schools? Where could GIS be used in schools? What does GIS look like in schools? How is GIS used in schools? Who uses GIS in schools? How are these things connected?

...

Take a minute or two to think before you start drawing.

Time allowed: 30 minutes. When 7–10 minutes before the end of the allocated time (or if participant is completed earlier):

Now before you give me your concept map (mind map) please take about 5 minutes to write a list of what you have drawn. Give additional sheets of paper as needed.

Thank you very much. You have done that really well. Please make sure your name is on your concept map and your list. Take a break and stretch as needed.

Take a moment to review what student/educator has drawn. Begin interview.

Data by participant

I will begin with middle school students' Ernest and CJ's representation and reflections. The representation shown in Figure 9.5 was the first, and only time Ernest used an image of any kind in one of his representations as most of his concept map representations were comprised of 'bubbles' of words with connections drawn between them.

When explaining the appeal of GIS to another student he notes, "Well, I'd probably say 'once you start using it, it's fun, because even if you may

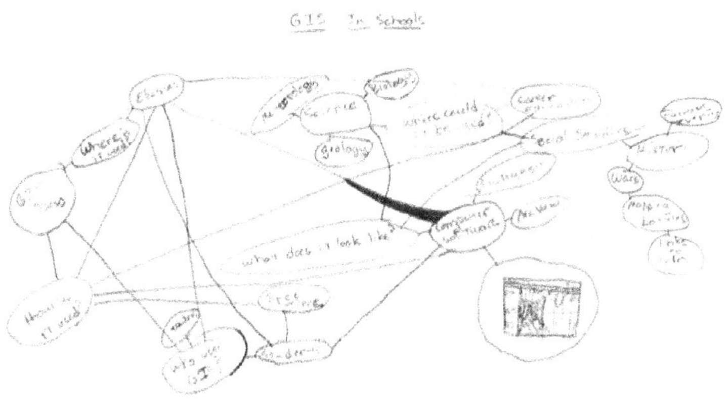

Figure 9.5. Ernest's representation of GIS in Schools (the only drawing of Ernest's to include an image.)

not be interested in GIS itself, you could use it in whatever career you are looking at. You can use it in your other classes. And it's fun."

Ernest included "GIS Day" as an important part of GIS in schools. GIS Day in Maple Middle School was celebrated for two years school-wide and one year on a 7th grade team.

Ernest described GIS day as:

> It was a day where each class would go down to the media center and the computer lab and we have stations. And they would just kind of rotate around the different stations learning how it's used where it's used...They would show you maps of things that they had made with GIS. The computer lab was used to teach a little bit about using the software and showed a couple of different features that might interest students in taking the class or being interested in it for a career.

CJ described it as an "instructional GIS day." He remembered some of the other guest speakers. In the computer lab, "she would have people finding their houses and stuff. We had a surveyor outside using a GPS system." During his second GIS Day celebration, CJ remembers helping in the computer lab. "I made sure everybody was able to do it in the computer lab. If they got lost, I helped them get caught back up."

The experiences students have with real world uses of GIS through projects can provide valuable services to the community while learning. Through events like GIS Day and other activities that invite professionals into the classroom and the school, many educators are raising the awareness of GIS and geospatial technologies.

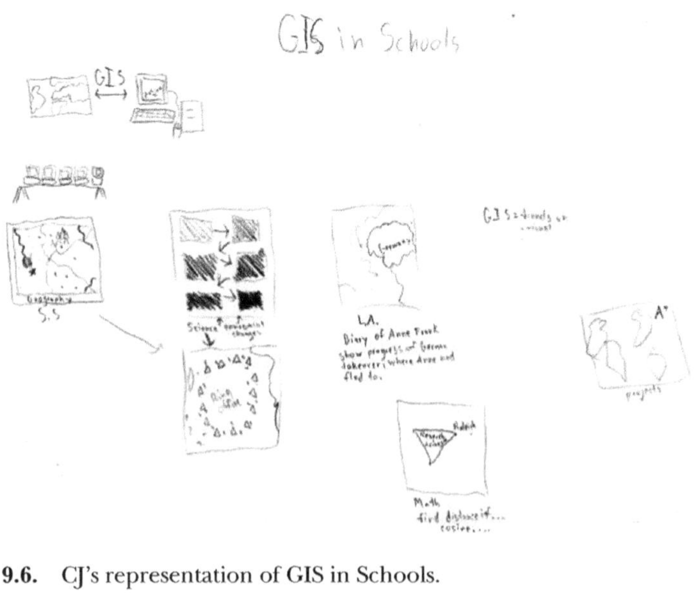

Figure 9.6. CJ's representation of GIS in Schools.

Discussion: Students

Learning style and personal interests of the students were explored as reasons for students to use GIS. Ernest mentioned the potential use of GIS in students' free time. Ernest had heard somewhere that students' "minds were like sponges, they just absorb everything." Ernest observed, "...it's easier for students to learn how to use the software because ... students aren't certain of a career path, so they're just interested in taking different classes and they're...it could easily spark the interest in someone to take more of a class, get more interest in it."

Students, as users of GIS in schools, were included in all 18 of the representational interviews and representations drawn by the participants. This is not to say that all students were seen as GIS users. Said Ernest:

> There would probably be some students who wouldn't like to use it because they don't like computers...I know some students would probably enjoy using it because it is something new and it's not just sitting in a class taking notes, it's doing something.

Gender preference as demonstrated by student enrollment in the GIS elective was also observed by the students at Maple Middle School. Ernest and CJ described their first GIS elective as having between 20 and 30 students, with more males than females enrolled. Both noted that their second level GIS class had 4 males only; no female students. Ernest commented:

> I don't know why. Part of it could be because the girls were drawn to more of the Home Ec classes and the guys are kind of drawn into technology and computer classes. So this was just another class that was based around computers and software.

Both students and educators commented on the grade level of the students who might use GIS. There was a hesitancy regarding the use of GIS in the lower grades. Ernest also felt that GIS might be overwhelming for the younger students and Natalie echoed CJ's thought that it could be used with the upper elementary students, if at all. CJ pointed out:

> I don't think it could be used for any grades that aren't introduced to computers yet. Less than third grade probably couldn't handle it. I think fourth and fifth would be good grades to be introduced to it. Get to understand it and know the tools.

He continued later stating, "I think you'd have to start GIS with third graders, fourth graders, fifth graders—students in elementary school—

learning what GIS is. And in middle school wanting to continue it—learning more about it."

The students stressed the importance of introducing GIS in later elementary grades and continuing to build skills through middle school and into high school. As these eighth graders were investigating different high school choices for the next school year, they had not heard of any high school using GIS. GIS was most often observed as a tool for students at a middle school or high school level by both students and most of the educators.

Discussion: Teachers

The observations and experiences of the participants in this study influenced their thoughts on the teachers as users of GIS. While all of the participants had a common understanding of the teacher as a user, they had very different observations about the characteristics of teachers who use GIS, the role of the teacher when using GIS and how teachers learn GIS.

Charles and Kathleen, teachers who had each taken an introduction to GIS course prior to taking the semester-long *GIS in Education* course, are discussed first. When describing the teacher in his representation, Charles remarked that he "is facilitating the learning of the software" (See Figure 9.7). Charles continued to describe the wetland area on his drawing that led to his personal feelings about his role as a teacher using GIS;

> I was thinking, a little bit selfishly, about myself because I don't feel like I like to be…tied to the desk. Not that I teach from a desk because everyone walks around a lot but just this whole paper/pencil—and that is important—but I think it would be good, healthy for us all to be able to get out more.

Charles' ideal, as a teacher, would be a combination of what he considers to be virtual and real experiences. As a participant in both his introductory week-long workshop and in the *GIS in Education* graduate course, Charles has been exposed to both hands-on applications and working with community partners, a requirement in the second course. As part of the rationale for the community partnership requirement, GIS in K–12 instruction was framed in a fourfold approach of *actual, virtual, critical* and *ethical* components of GIS education.

As Kathleen began to consider the opportunities necessary to raise awareness of GIS in schools she raised the first of her two questions found on her representation: "What is the rate of growth of GIS in schools?" (see Figure 9.8). She remarked during the interview, "I really don't have a sense of that—How big the community is." Her initial observations were based on her subscription to the EdGIS listserv (2001–2005) (https://

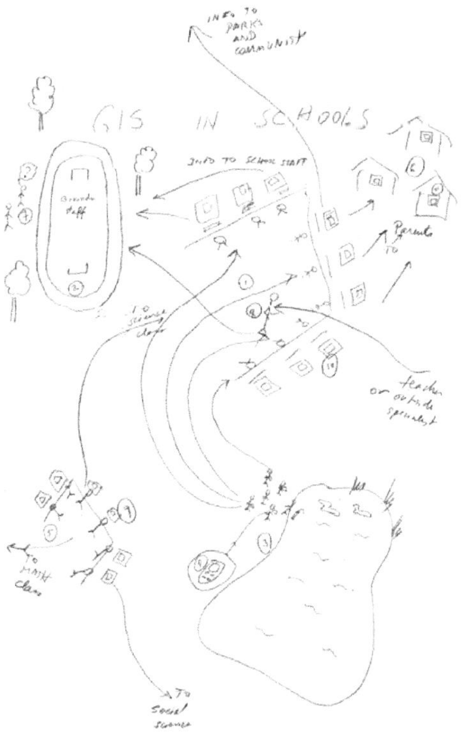

Figure 9.7. Charles' representation of GIS in Schools.

list.terc.edu/mailman/listinfo/edgis). She observed that for a long period of time there were very few posts. She noted there seemed only to be one individual, who was nationally recognized, that posted announcements regularly to the list. More recently, she observed, other individuals posting questions to the list:

> So, I'm thinking there must be more people who are participating and…but it's still sort of the individual teacher who thinks, "Oh this is really cool, I'm going to use this in my class" and the kids get exposed to it that way.

Kathleen's train of thought began with the question: "What's the rate of growth? And is it the kind of thing where if one teacher gets hip to it then it just starts taking over?" She then immediately wondered, "In what community has GIS been fully integrated in to schools and instruction?"

The question of "the rate of GIS" that Kathleen raises is a question of the DI s-curve (see Figure 9.1) as well as the adopter categories (see Figure 9.2). GIS adoption in K–12 schools is still in the initial stages, as shown in

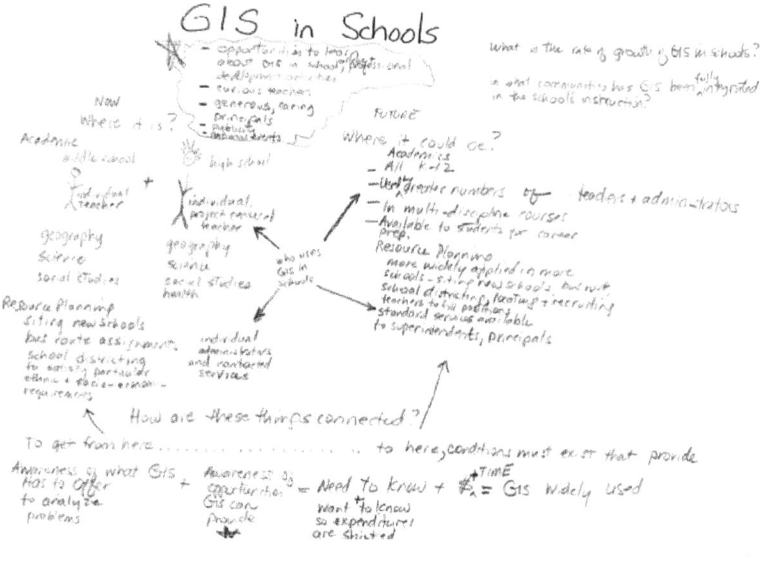

Figure 9.8. Natalie's representation of GIS in Schools.

Figure 9.4, the overlay of the innovation curve with the organizational decision process.

The adoption of GIS, at a slower rate therefore falls into Rogers' initial stages of *agenda setting* and *matching*. The K–12 GIS adopters are at his '*early adopter*' stage, only a small percentage of the teacher population. The *innovators* and *early adopters* in the GIS in education field are a major factor influencing the rate of adoption of GIS in schools. The software manufacturers, or change agents in this process, are developing online GIS data and online Java-based products that do not require as much of the hardware, and reduce the need for purchased software or trained teachers. These accommodations are intended to influence the adoption rate as well.

Natalie raised a similar question to Kathleen's, regarding adoption, "What schools are using it now?" This curiosity about others using GIS and how they are using it has implications for the businesses and organizations hosting GIS in education listservs and creating professional development opportunities in GIS. These data suggest the need for a community of users in the educational fields and opportunities where educators can learn about others who are in similar situations. While there are numerous websites, conferences and listservs geared toward GIS in education, there are GIS users who are not aware of what is available to them. Natalie's question lends itself to a larger discussion of the awareness of GIS use in schools and the awareness of the potential of GIS in schools.

Kathleen's image is divided into three parts (see Figure 9.9). The left of the image demonstrates her understanding of GIS in schools as she sees it

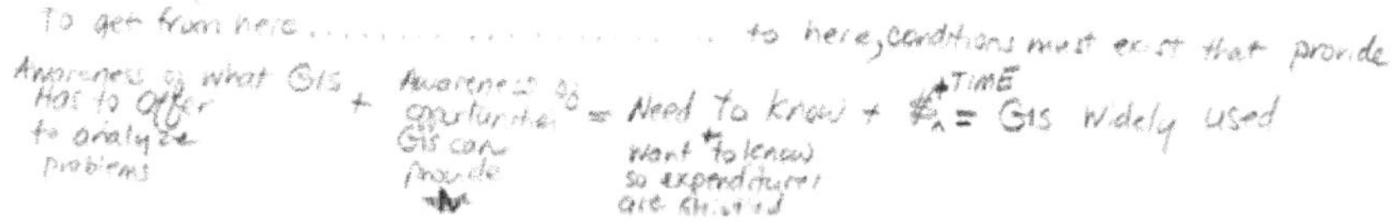

Figure 9.9. Kathleen's formula for widespread implementation of GIS in schools. This figure is part of Kathleen's larger representation of GIS in schools.

now, in the present. The right side of the image demonstrates her understanding of GIS in schools in the future. The final part of her representation is located along the bottom of the drawing and it is her illustration of how "to get from now to the future." The starred cloud at the top of the representation is her brainstormed list of how to "raise awareness of opportunities GIS can provide" found at the bottom of the image.

The formula as shown in Figure 9.9 suggests what Kathleen describes must be in place for the future implementation of GIS in schools. Kathleen described her personal understanding:

> On a personal level I just thought of the awareness of what it has to offer and how to make people aware. And not only what it has to offer but opportunities—the problems it can solve. Once you know that you need to know this stuff, then you turn around and shift expenditures so that you can find the money and a time to get it implemented.

This conceptualization of how to move forward in the future with GIS in schools is similar to the *Agenda-Setting* stage of Rogers' (2003) innovation process. Within this first step, organizations often seek out an innovation in order to solve a known problem. However, Rogers recognized that "Sometimes knowledge of an innovation, rather than the recognition of a problem or need by an organization leading to search for a solution, launches the innovation process" (2003, p. 423).

Community Partnerships in the Agenda-Setting Process

As part of Kathleen's agenda-setting to help her school embrace GIS, she procured school funding to bring a national speaker to visit and to broadcast his talk through her distance education network. Kathleen described the importance of one part of the visit to her school from a noted national GIS professional:

> We did these presentations when [he] was here where he just talked about what GIS was. And I think that that is part of his mission to just get the word out there so he showed examples of where it's used and how it's used. He gave it a name, which is what I think, is partly what the problem is. People

> can't understand it unless they know that there is this body of software and technology. Oh, yes, you've used MapQuest—you've used GIS.

The idea of the community as a partner in GIS in schools was suggested in many of the educators' and students' interviews. Community partnership was modeled from the earliest *GIS in Education* course through guest speakers. The student participants also were encouraged to establish connections to the community through the juvenile crime data project they created in the second level GIS elective. They also participated in, and described GIS Day celebrations where GIS professionals were invited to the school.

Course and Curriculum Areas

Building upon the concept of the teacher as curriculum expert, Natalie provides a way to begin to introduce GIS to classroom teachers:

> Teachers love curriculum, so if we back it in—technology in their curriculum—instead of saying ok we're going to learn this piece of software, GIS...You take a piece of curriculum and say this is the curriculum—What can we do? How can we enhance it? And start from there, because they know the curriculum. And then stuff it in after the fact. That's what we do. We say here's the technology, learn how to use it. Ok now, what can we do with this? Nothing, it takes too much effort. So if you start with the curriculum and work your way up...

Natalie's image shows GIS as fully integrated across the curriculum (See Figure 9.10). Like Kathleen, Natalie was in an "other-than-classroom" role in her school. As a middle school Curriculum Integration Coordinator, her role was to locate, develop and provide cross-curricular connections for middle school teachers and learners. Thus, to her, GIS was an excellent vehicle for integrating these subjects. As a doctoral student in the Instructional Technology program, Natalie was strong on technology skills, which facilitated keeping her focus on the curriculum as she learned GIS. For the *GIS in Education* course, Natalie developed a "Techno Book Review" based on the required 8th grade reading list of journey books. She designed a cross-curricular assignment of mapping the journey with GIS, calculating the mileage and estimating the costs in math, and for science, describing the seasonal climatic conditions of the journey (Alibrandi, 2002).

One problem that many K–12 educational institutions face is the sustainability issue that Rogers (2003) noted. Many of the individual schools make a decision to purchase the software based on the use of GIS by the early adopter in their school. If this pivotal individual leaves the school what happens? This individual may influence the adoption of the technol-

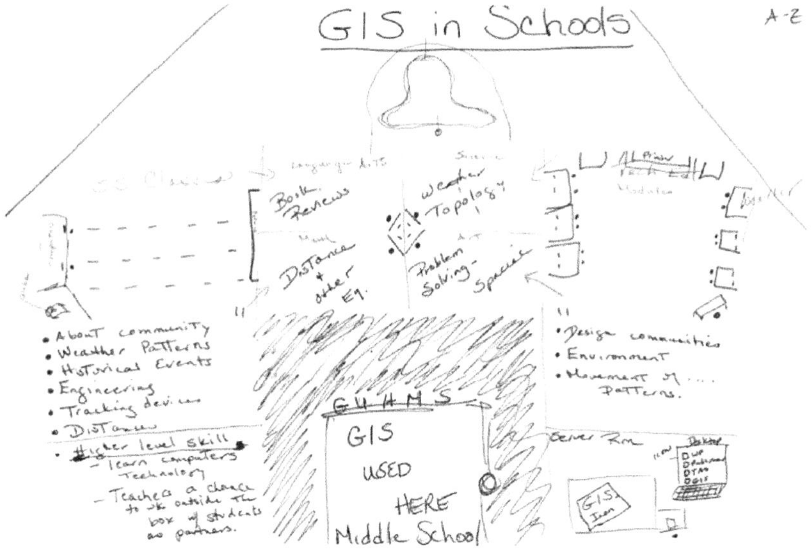

Figure 9.10. Natalie's representation of GIS in Schools. She began with the superstructure of the school and named it GUHMS—GIS Used Here Middle School. She explained, "GIS in this particular case would be within…this is the access we have within our walls."

ogy by others in the school, as Charles had through in-service opportunities offered to his colleagues. That person becomes the support for the other learners and if that pivotal individual leaves the school, the question among many researchers is, what happens to GIS in that setting?

The students observed this set of circumstances at Maple Middle. When Ernest was asked about why he thought his school decided to have a GIS elective he responded:

> I think it was a teacher influence thing. The teacher that taught the class, it kind of came and went with her. And there were a couple of other teachers, I think, that had seen it used before and were kind of pushing the school into creating an elective for it. And then GIS evolved from that.

CJ provided the following reasons why he felt the electives were no longer available:

> There's no one with enough knowledge on the subject. I don't think anyone at Maple anymore has learned about GIS, Mr. Swan left, Mrs. King left.

Unfortunately, students are most likely to lose opportunities when GIS is not sustained in an educational institution. Ernest pointed out during the interview that the GIS elective was no longer on Maple's list of electives.

When asked why it was no longer on the list, he replied, "Well, I don't know why. I really wish it was still there because it was a really cool class. It's neat software that I enjoyed using. I don't know why we don't have it anymore."

Alibrandi (2003) has written about the 'love/hate relationship' between education and innovation—the many cultural icons of learning and teaching that prove resistant to change, and the multiple factors that limit integrating spatial content and technology applications. While the *GIS in Education* course that was common to the educators' experiences focused on GIS as an interdisciplinary application, the inevitable adherence to traditional curriculum and methods persists. Most discussions of GIS in K–12 have chronicled the slow rate of adoption, but under time and testing pressures teachers would have to believe that GIS would make instruction *easier* in order to consider adoption, and until the Internet-based GIS tools appeared, that wasn't the case. This simply underscores the importance of the s-shaped adoption curve reflected by K–12 Internet integration.

Ben, a social studies teacher, discussed the pitfalls of chronologically organized textbooks in social studies. He felt that this provided an opening for GIS:

> You learn about this time period and you're skipping from this part of the world—maybe Asia—to what's going on in South America at the same time. So the students do not get a really fair representation mentally. You show it on a map but there's only so much a map does. But you show them something on a computer where you can pull two aspects of the world apart and you can show them similarities and differences.

Uncertainty

Ben's representation began with a giant question mark with the words "Where," "Who," "When," "Why," and "How" (see Figure 9.11). He related these questions to his thoughts about individual teachers using GIS:

> Why would a teacher use it? That is pretty simple—to help students learn. It's kind of helped my knowledge as well. That's kind of set. The how—it depends on the teacher. If you're going to use GIS at all...integrate it into the classroom. That's their choice. Every teacher is different. When, when will it be in your curriculum? And who?...

Ben continued to consider students as the 'who,' but he also elaborated that his question mark is representative of many more questions. He explained it was more than just one teacher asking these questions. He felt that parents, teachers and school systems should play a role in this questioning process.

Figure 9.11. Ben's representation of GIS in Schools.

Ben's uncertainty surrounding the decision making process of GIS adoption is indicative of the matching stage of DI theory. Before any decision can be made, an agenda must first be set and then the organization and its stakeholders must begin to match the needs of the organization with what the innovation offers. Another point of uncertainty that Ben demonstrates in his representation is the sign that reads:

GIS
Not Here
Be Back @ __?__

Ben stated, "I used the sign to illustrate 'Not here today, Be back at ?'… When is it going to be here, if it's not here already?"

If the decision-making process is rushed it is possible that the innovation will not succeed, even if there is a desire for the adoption of it. Rogers (2003) points out the importance of the matching stage, "Effectively matching an innovation with an organization's need is key to whether the new idea is sustained over time" (p. 423).

SUMMARY OF FINDINGS

What is presented in this chapter is only a fraction of the 18 interviews and representations from the research (White, 2005), the following synthesizes the findings presented herein. The common categories of description that emerged surrounding the investigation of educator and student perceptions of GIS in schools were clearly defined. They included the categories of description: a) GIS technology in schools, b) users of GIS in schools, c) classes and curriculum areas where GIS could be found/used, d) examples of potential uses of GIS in schools, and e) awareness issues.

Areas of convergence were found between GIS technology, the users of GIS, GIS in classes and curriculum areas, and the potential uses of GIS in schools. From this murky area, a final category of description emerged: e) awareness issues. Uncertainty seems to envelop issues that are directly connected with the innovation decision process whether or not GIS is an innovation worthy of the effort and able to meet the needs of the stakeholders. The questions raised by the participants regarding the rate of growth, the influence of individual teachers in the K–12 setting, and the larger GIS in education community, are also influenced by the DI process. These questions correspond to the rate of diffusion, the adopter categories, and the stages of adoption within school organizational structures respectively.

One of the contributing factors regarding GIS awareness is a lack of name recognition. While most TV viewers can intuitively interpret a TV weather map, it is not called a GIS. Therefore, GIS has an identity problem—it is still invisible to the degree that the many constituent applications people see, use and even manipulate daily are not referred to as GIS. If a technology or any innovation does not have a name, even if it is ubiquitous, it cannot be set apart and seen as anything other than what an individual understands it to be. Once it has name recognition, GIS becomes a discrete entity that can be described and understood more fully.

The student and educator conceptualizations of GIS in school revealed that they located themselves from both allocentric and egocentric positions in relation to GIS. The categories of description that were discussed from an allocentric position were: the GIS technology in schools and the classes and curriculum in which GIS could be found. The participants were able to discuss these concepts but placed a distance between the idea in schools and themselves. The egoentrically positioned categories of description that emerged were: the users of the GIS technology in the school and the uses of the technology in curriculum and subject areas.

This point of confluence however did produce a turbid fifth category. The final category of understanding of the topic of GIS in schools was: awareness issues that emerged through uncertainty and questions posed by the participants. As the educators in this study struggled with the topic of GIS in schools, they had grey areas that emerged in their own understandings. They asked questions both in the representations and in the interviews. The significance of the quandaries and question marks is an indicator of the stage of diffusion of GIS in schools in Rogers' matrix. GIS in schools is located in the initial stages of diffusion. In an attempt to set an agenda and match organizational goals, these stakeholders are asking questions to shift toward the implementation of GIS in schools. Figure 9.12 helps to illustrate the relationships and outcomes of the "pools" of common understanding as they pertain to GIS in schools.

Common understandings of the participants in this study regarding GIS in the world and GIS in schools, were found in the technology and in the community. While there were participants that did describe GIS within the context of instructional technology, it was not commonly understood by all participants as such. The connection between how GIS was viewed in the world and how GIS was viewed in the school as a distinct technology was more commonly understood. GIS was seen as software with distinct characteristics that can be set apart from other IT software packages available. The ability to visualize tabular data in geographic and spatial modes is a distinguishing feature of the application.

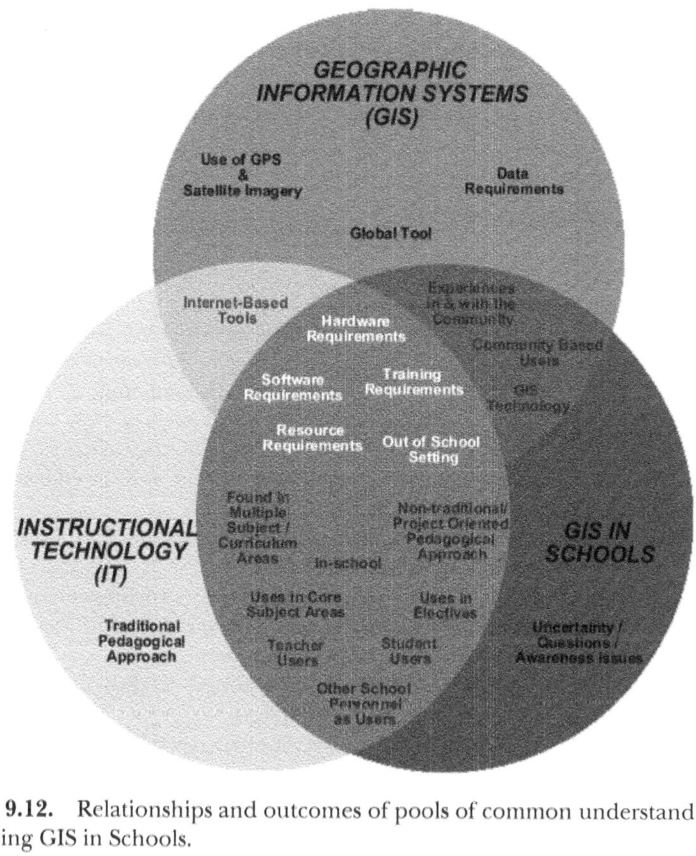

Figure 9.12. Relationships and outcomes of pools of common understanding regarding GIS in Schools.

The other categories of understanding with commonalities between GIS in the world and in the school were in the realm of community experiences and community based users. The educators and students were able to identify community based problems and examples of GIS use for the classroom. In addition, all participants described the role of community partners in the classroom, which may have been influenced by the context in which these participants were taught GIS.

White's (2005) "areas of turbidity" are the locations of important tensions that still influence K–12 GIS adoption. The value of applying DI theory, particularly the matching process, may be viewed as the current location and status of the producer/consumer tension, this research and indeed, this volume. Therefore, the utility of DI analysis in viewing GIS adoption can be helpful as part of a continuing research design.

As the landscapes and contexts of the Internet, GIS, IGIS and IT in schools continues to change with evolving hardware (handhelds) and software, the matching stages of adoption should accelerate the s-shaped adoption curve. But the identified questions of how, when, where and by whom remain the inevitable tension in that process. Others in this book are addressing these tensions from a variety of angles—the industry angle; the curricular angle; the teacher education angle; the Internet and software development angles, and so on. These collective efforts, which may all be seen within a DI model, are converging in Rogers' matching process.

REFERENCES

Alibrandi, M. (2002). Geography is everywhere: Connecting schools and communities with GIS. *Learning and Leading with Technology, 29*(7), 32–37.

Alibrandi, M. (2003). *GIS in the classroom: Using geographic information systems in social studies and environmental science.* Portsmouth, NH: Heinemann.

Ausubel, D. P. (1963). *The psychology of meaningful verbal learning.* New York: Grune and Stratton.

Baker, T. R. (2005). Internet-based GIS mapping in support of K–12 education. *The Professional Geographer, 57*(1), 44–50.

Bogdan, R., & Biklen, S. K. (1998). *Qualitative research for education: an introduction to theory and methods.* Boston: Allyn and Bacon.

Dall'Alba, G. (1996). Reflections on phenomenography: An introduction. In G. Dall'Alba & B. Hasselgren (Eds.), *Reflections on Phenomenography: Toward a Methodology?* Göteborg, Sweden: ACTA Universitatis Goth

EDGIS Listserv. (2001–2005). TERC. Messages posted to EdGIS mailing list, archived at http://list.terc.edu/pipermail/edgis.

Environmental Systems Research Institute (ESRI). (2004). *History of ESRI.* Retrieved June 23, 2004.

Environmental Systems Research Institute (ESRI). (2004). *ArcExplorer—GIS Data Viewer.* Retrieved December 3, 2004, from http://www.esri.com/software/arcexplorer/index.html

International Society for Technology Education (ISTE). (2000). *National Educational Technology Standards for Students: Connecting Curriculum and Technology.* Eugene, OR: The Society.

Kerski, J. (2003). The Implementation and Effectiveness of Geographic Information Systems Technology and Methods in Secondary Education. *Journal of Geography, 102*(3), 128–137.

Kerski, J. J. P. (2000). *The Implementation and Effectiveness of Geographic Information Systems Technology and Methods in Secondary Education.* Unpublished Ph.D.

Lynch, K. (1960). *The Image of the City.* Cambridge, MA: Technology Press.

Malone, L., Palmer, A. M., & Voigt, C. L. (2002; 2005). *Mapping our world: GIS lessons for educators.* Redlands, Calif.: ESRI Press.

Marton, F. (1981). Phenomenography: Describing Conceptions of the World Around Us. *Instructional Science, 10,* 177–200.

Marton, F., & Booth, S. (1986). Phenomenography: A research approach to investigating different understandings of reality. *Journal of Thought, 21,* 28–49.

Marton, F. (1994). Phenomenography. In T. Husen & T. N. Postlethwaite (Eds.), *International Encyclopaedia of Education.* Oxford: Pergamon Press.

Marton, F., & Booth, S. (1997). *Learning and awareness.* Mahwah, N.J.: L. Erlbaum Associates.

Marton, F., & Fai, P. M. (1999, August 24–28). *Two Faces of Variation.* Paper presented at the 8th European Conference for Learning and Instruction, Göteborg University, Göteborg, Sweden.

Mavers, D., Somekh, B., & Restorick, J. (2002). Interpreting the Externalized Images of Pupils' Conceptions of ICT: Methods for the Analysis of Concept Maps. *Computers and Education, 38,* 187–207.

National Center for Educational Statistics (NCES). (2004). *NCES Education Data Search Tools.* Retrieved March 27, 2004 from http://nces.ed.gov/datatools/.

Novak, J. D., & Gowin, D. B. (1984). *Learning how to learn.* New York: Cambridge University Press.

Paivio, A. (1986). *Mental representations: A dual coding approach.* New York: Oxford University Press.

Paivio, A. (1991). *Images in mind : the evolution of a theory.* New York: Harvester Wheatsheaf.

Rogers, E. M. (2003). *Diffusion of innovations* (5th ed.). New York: Free Press.

Somekh, B., & Mavers, D. (2003). Mapping Learning Potential: students' conceptions of ICT in their world. *Assessment in Education, 10*(3), 409–420.

Surry, D. W., & Ely, D. P. (1999). *Adoption, diffusion and implementation and institutionalization of educational technology.* Retrieved February 15, 2004, from http://www.coe.usouthal.edu/faculty/dsurry/papers/adoption/chap.htm

Vosniadou, S., & Brewer, W. F. (1992). Mental Models of the Earth: A Study of Conceptual Change in Childhood. *Cognitive Psychology, 24,* 535–585.

White, S. (2005). Geographic Information Systems (GIS) and Instructional Technology (IT) Diffusion: K–12 Student and Educator Conceptualizations. Unpublished Ph.D. dissertation, North Carolina State University. Retrieved April 23, 2007, from http://www.lib.ncsu.edu/theses/available/etd-09092005-050901/

CHAPTER 10

GOING FLAT

The Changing Dynamics and Integrative Nature of Geography in the Digital Classroom

Jay D. Gatrell
Indiana State University

John L. Conant
Indiana State University

The publication of Thomas Friedman's (2005) *The World is Flat: A Brief History of the 21st Century* has transformed the way policy makers, the public, and even academics understand and know the world around them. The public salience of this book has been closely aligned to the mounting realities of globalization that impact the daily lives of people and places in the U.S. and around the world. Consequently, the notion of the world as a flat isotropic plane without natural borders such as mountains, rivers, and oceans or governmental borders such as regulations or restrictions on the movements of capital, people, and finances alters the world view of individuals as they struggle to make sense of how places change and time-space compression increases and intensifies the competition between places, states, and nations for valuable resources. But for geographers and more importantly social studies teachers, Friedman's book is important for two

Digital Geography: Geospatial Technologies in the Social Studies Classroom, pages 197–211
Copyright © 2008 by Information Age Publishing
All rights of reproduction in any form reserved.

reasons. First and foremost, the public discourse on globalization has heightened—or at-least has the potential to heighten—the importance of social science education and the relevance of geography to the social studies curriculum. Second, Friedman's book proves especially problematic in that it is simultaneously global and local by reducing the world to "supply chains" and extra-territorial flows of human and finance capital across the borderless space of globalization. The purpose of this chapter is to examine how these new intersections complicate the more abstract notions of "space" and "place" within the discipline of geography. In this chapter, we will discuss the importance of considering how academic geographers have thought about space and place within the context of the National Geographic Society's Five Themes of Geography. Next, we will examine the ways in which cartographic representations complicate the teaching of space and place, and examine one innovative curricular strategy for teaching how space and place have changed over time. Thus, our chapter will sketch a framework for the use of digital resources that are disciplinarily appropriate, socially relevant, and leverage existing content standards and curricula to infuse geography throughout the social studies curriculum. As the future competitiveness of the U.S. economy requires the institutionalization of a new integrative and collaborative approach towards K–12 education, the social studies curriculum is itself globalizing and creating new connections between previously independent fields of study. Just as the social and economic walls between nations have come down, the traditional walls isolating the social science disciplines are being breached as the realities of globalization infiltrate the K–12 curriculum.

ENGAGING THE DISCIPLINE, MAKING IT RELEVANT, AND THE STANDARDS: HOW DO GEOGRAPHERS GLOBALIZE SPACE AND PLACE?

One of the most salient critiques of the geography education community has long been that geography education in the K–12 environment seldom reflects the realities of the post-secondary curriculum (Gatrell, 2004; Warf, 1999). Despite the historical cleavage that has existed between the two communities, the emergence of globalization as a major theme in contemporary social studies curriculum and more importantly curricular reform initiatives as outlined in the National Academies' (2005) report "Rising Above the Gathering Storm" has afforded geography education the unique opportunity to leverage the existing and vast geographic literature on globalization based in the post-secondary literature. In many respects, the new globalized social studies curriculum that is evolving mirrors the historical framework of the middle school curriculum and is resulting in a more holistic and integrative account of social science at the high school level, too.

One of the most developed literatures related to globalization examines the local-global nexus and how globalization is expressed in local places. That is to say, academic geographers are interested in how place changes in response to global socio-spatial processes and how local processes may (or may not) inform global ones. The dominant framework for understanding how local places experience globalization is sometimes referred to as the place-space dialectic (or local-global dialectic) (Gatrell & Collins-Kreiner, 2006). The dialectic has been structured in such a fashion that place refers to local realities and space is the site of global processes (McCann, 1999; Merrifield, 1993a, 1993b). And it is through the very mechanisms (e.g., telecommunications); structures (e.g., global capitalism), institutions (e.g., World Trade Organization and World Bank), and processes (e.g., the international division of labor and collaboration production) of globalization that the geographies of communities change. Accordingly, geographic research has used the dialectic—as well as Henri Lefebvre's (1991) spatial practices (including monetary policies, deregulation/regulation of industries, and flexible labor)—to understand how global flows of capital and labor translate into altered everyday realities in local places (see Figure 10.1). In short, academic geographers are intrigued by how the imagined spaces of globalization alter everyday places. Within the context of the middle school and high school curriculum, space and place can be used to demonstrate the uncertainties and familiarity of students with globalization and their own communities, respectively.

For many K–12 students, the realities of globalization are decidedly concrete in that globalization—*writ large*—may be personally experienced through industrial restructuring and job loss (or as is more often the case but less often recognized, job creation). Alternately, globalization may be understood as a virtual reality that enables children to communicate online with peers from around the world. Just as likely for the K–12 student, globalization may also be a little known process that is most concretely understood through the UPS "Brown" global supply chain commercials or the South Asian accent on the other end of the phone when calling a phone support center. In any event, it is evident from these diverse examples that globalization is an ambiguous and confusing process

Figure 10.1. The Local-Global Dialectic based on Lefebvre (1991).

that does not fit neatly within any single curricular framework per se. Yet, the possibility exists that geography educators—and social studies teachers broadly—can use the themes of space and place to explore globalization and teach content across the social studies curriculum including history and economics using digitally accessible map libraries or other internet resources.

In practical terms, the geographies of space and place can most readily be understood as the disjuncture that exists between the digital and analog worlds. Or to use the language of the National Geographic Society's Five Themes—"location" (i.e., locating a place in space) does little to provide students with a sense of "place." While no doubt location and place are similar concepts—they are qualitatively distinct and the emergence of geospatial technologies (GST) has the potential to displace the everyday lived experiences and materiality of place and substitute the simulacrum and novelty of GPS coordinates and Google Earth. As such, the challenge for geography in a digital classroom is to provide students with a sense of place—while leveraging the benefits of technology in the classroom. In the process of conveying a sense of a place, geography educators can integrate GST across the curriculum.

One example of how space and place have changed everyday life over time and across space is the location and structure of markets. Historically, markets have been market-places. Today, markets exist in the hyper-space of globalization and the internet. Instead of going to the mall, consumers now access virtual market-spaces, such as eBay, Amazon, or even Wal-Mart. Likewise, workplaces have been transformed into workspaces. Indeed, the language of place has been transformed into space in almost every aspect of daily life. Space has become a platform for activity and it is through the lived experiences that featureless spaces become meaningful places. The retooling of space and place in the public discourse has come full circle and has even transformed the concept of a house—as a space (think of the television show *Trading Spaces*)—and home as a place. In the end, the abstract notion of space and place are considerably more concrete than "academic" geography and can be used to inform the social studies curriculum.

ALTERED GEOGRAPHIES: HOW HAS DIGITAL GEOGRAPHY CHANGED THE LANDSCAPE?

Social studies teachers should strive to use GST to infuse geography throughout the curriculum. As such, the suite of GST should be broadly conceptualized to include a full range of digital resources including GIS, remote sensing, and global positioning systems—as well as map archives, photograph libraries, and digitized art. Indeed, the availability of digital resources is so vast that technology has the potential to be the primary determinant of the learning environment; thereby potentially diminishing, albeit unintentionally, the core disciplinary outcomes and standards (Gatrell, 2001, 2004). However, technology—when positioned within an appropriate conceptual framework and anchored by standards has the potential to unlock geography's inherently integrative nature and expand access to geographic knowledge. As part of these integrative trends, the National Council on Economic Education (NCEE) and the National Council for Geography Education (NCGE) have even developed curriculum pieces that are explicitly integrative across traditional course boundaries (e.g., NCEE's *Middle School World Geography: Focus on Economics* and *Geography: Focus on Economics*).

In the previous paragraphs, we alluded to the use of digital map libraries and Google Earth to understand how geographies change and how the digital classroom and globalization may alter the perceived geography of communities. These resources—while not examples of GIS or remote sensing per se—are fine examples of how digital resources—writ large—can enhance the geography classroom. Indeed, these resources can be used to introduce concrete geographic principles and increase student awareness of GIS and remote sensing through basic manipulation of remote sensing data. To that end, examples are presented in Maps 10.1 and 10.2.

The differing cartographic representations of the analog and digital worlds (to be read place and space) illustrate in concrete terms the role technology has played in transforming our collective understanding of

Map 10.1. Panoramic Representation of Terre Haute, Indiana, 1880. *Source:* Library of Congress, Geography and Map Division, http://memory.loc.gov/, Downloaded August 8, 2006.

Map 10.2. Terre Haute, Indiana, 2005. *Source:* Google Earth, 2006. Image accessed August 10, 2006.

local social systems. In many respects, Google Earth (and similar websites such as Microsoft's TerraServer and the USGS's many state-based image servers) has become the contemporary version of the panoramic map. Unlike the panoramic map, Google Earth allows the student to "jump" scale and manipulate digital data including GIS layers such as roads and place names. Yet, the character and sense of place conveyed by each is distinct. Whereas the digital imagery of satellites affords an interesting perspective and allows the reader to rapidly jump from the global to the local with the click of a mouse, the 1880s representation provides the map reader with a unique and historically distinct perspective on a place. In the process, the panoramic view provides the reader with insights into the topography and physical geography of the region in a fashion unavailable to the Google Earth viewer as the Google image is literally flat when the "tilt" tool is adjusted by the student (Map 10.3).

In concert, both images can be used to explore how the themes of geography have changed as cities evolved from places of production with distinct identities to spaces of production connected by highways and now virtual networks. For example, how has "movement" of people, products, and ideas changed over time? What are the dominant features of each map? Why are they different in each? How has technology transformed the landscape of each? How have "human-environment interactions" changed

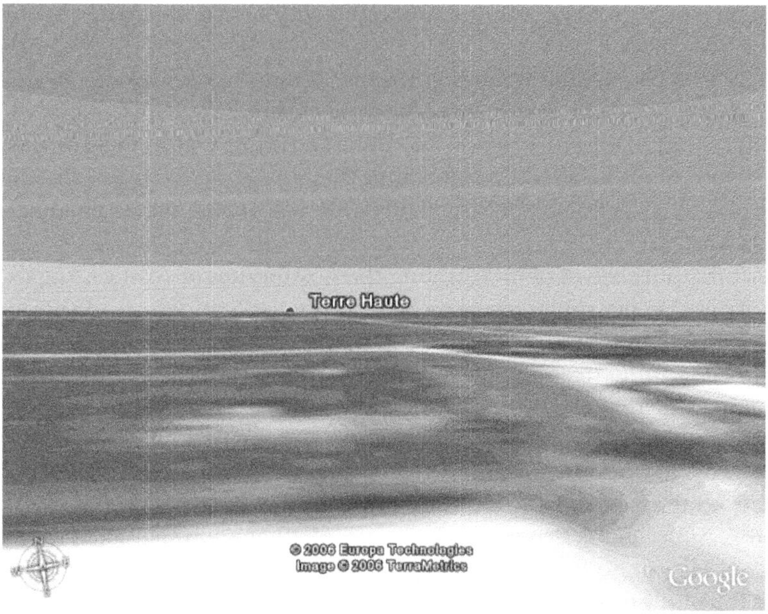

Map 10.3. The World is Flat? *Source:* Google Earth, 2006. Image accessed August 10, 2006.

over time in Terre Haute? Clearly, digital resources enable teachers to ask interesting questions above and beyond where things are and allow geography to transcend disciplinary (or curricular) boundaries. In the next two sections, two examples are provided. The first example—Geography & History of the World—illustrates how the State of Indiana's Core40 curriculum[1] has been retooled to include an inter-disciplinary course that uses the concepts and tools of geography to understand globalization from 1400 to today. The second example is derived from a series of workshops sponsored by the National Council for Economic Education that explored the intersections between economics, geography, and governance issues as expressed through the globalization process.

HISTORY: GEOGRAPHY & HISTORY OF THE WORLD

Indiana's Geography & History of the World "power standards" and individual "indicators" include specific references to space and place, as well as global change. Not surprisingly, the standards place a major emphasis on issues of cartographic representation, and by implication GST. In this respect, history becomes the vehicle for understanding spatial processes. The course is unique in that it does not mandate a specific chronological sequence or regional account. Rather, the emphasis is on change over time and across space.

The new course's integrative curricular framework is ideal for investigating themes of place and space. Also, the historical change element will enable social studies instructors to combine primary sources—such as diaries—with maps to obtain insights into the how places changed and movement of ideas, colonial infrastructures, or nationalist ideals changed the landscape.

For the purpose of demonstrating the possibilities of synthesizing digital history and geography resources, we will discuss Standard 6—Innovations and Revolutions: Students will examine physical and human geographic factors that influenced the origins, major events, diffusion, and global consequences of new ideas in agriculture, science, culture, politics, industry, and technology. As might be expected, one of the main indicators, or substandards, focuses on the industrial revolution. To that end, teachers can access a variety of digital resources, such as Fordham University's Internet History Sourcebook Project [http://www.fordham.edu/halsall/] managed by Paul Halsall, to locate and identify public domain texts. Halsall's Industrial Revolution materials include first person accounts of the changes in the daily lives of workers and the materiality of their communities. One example of this is Engels (1892) 1844 account of "Industrial Manchester." The excerpt is prefaced by a explanatory note that includes the general geography of the region and how Manchester came to be a major site during the Industrial Revolution—indeed perhaps the first truly industrial city. See an excerpt of Engels (1892) below:

> Manchester lies at the foot of the southern slope of a range of hills, which stretch hither from Oldham, their last peak, Kersall moor, being at once the racecourse and the Mons Sacer of Manchester. Manchester proper lies on the left bank of the Irwell, between that stream and the two smaller ones, the Irk and the Medlock, which here empty into the Irwell. On the left bank of the Irwell, bounded by a sharp curve of the river, lies Salford, and farther westward Pendleton; northward from the Irwell lie Upper and Lower Broughton; northward of the Irk, Cheetham Hill; south of the Medlock lies Hulme; farther east Chorlton on Medlock; still farther, pretty well to the east of Manchester, Ardwick. The whole assemblage of buildings is commonly called Manchester, and contains about four hundred thousand inhabitants, rather more than less. The town itself is peculiarly built, so that a person may live in it for years, and go in and out daily without coming into contact with a working-people's quarter or even with workers, that is, so long as he confines himself to his business or to pleasure walks. This arises chiefly from the fact, that by unconscious tacit agreement, as well as with outspoken conscious determination, the working people's quarters are sharply separated from the sections of the city reserved for the middle-class; ...(p. 45)

Figure 10.2. *Over London by Rail* by Gustav Lore, c1870. *Source:* Public Domain from Wikipedia.org.

The availability of rich historical texts like one above will enable students to understand the places that populated the space of industrialization in the 1800s. Other online resources include a letter to the editor describing the "filth on the Thames", as well as artwork by Gustav Dore (Figure 10.2) that can be used to explore how the Industrial Revolution altered "human environment interactions" (Standard 9) or the "urban landscape" (Standard 5); thereby meeting or reinforcing related course objectives. The Engels example or the Dore print when combined with a map of the Industrial England enables students to discuss a wide array of complex issues including technology, economics, human environment interaction, and class (see Map 10.4). In the case of the new Geography & History of the World course, the digital classroom and the availability of GST—broadly conceptualized—have the potential to infuse geography throughout the curriculum. In the next section, we will discuss points of potential integration within the economic curriculum.

ECONOMICS: WHY REGIONS MATTER?

High school economics is one of the high needs content areas around the nation and an increasingly important component of the contemporary curriculum and policy debates surrounding the preparedness of high school graduates to compete in a global society. In Indiana, economics is required for the Core40 and high school honors diploma. However, the availability of licensed economics teachers and, to a lesser extent, geography teachers is limited. To meet the need for both licensure areas the Center for Economics Education and Social Science Education Center, along with the

Map 10.4. Industrial England since 1750. *Source:* University of Texas, Perry-Castaneda Collection [http://www.lib.utexas.edu/]. From W. Shepard (1991) *The Historical Atlas.*

Indiana Department of Education, partnered to create a series of professional development opportunities under the rubric of globalization funded by the National Council on Economic Education.

To meet the growing need for economics and geography licensure and to update (read globalize) the skill set of practicing economics, geography, and government teachers, a series of three inter-related workshops was envisioned. The workshops were multiple day residential workshops and trained between 13–15 in-service and 2 pre-service teachers per year. The workshop was targeted to key teachers with leadership roles, such as department chairs and curriculum coordinators, to leverage the curricular impact at the individual and school corporation scales. The objective of each workshop was to equip faculty with up-to-date knowledge and strategies for making the curriculum relevant to the daily lives of students through the use of popular narratives on globalization and to provide faculty with a range of teaching materials including hands-on lesson/activity development and the opportunity to reflect on their own classroom practice.

During each phase of the globalization series, the objective was to alter the scale and scope of the workshops and to expand the curricular target each year. The 2005 workshop was organized around the geography of global economics and emphasized international trade. The center piece for Year 1 was Thomas Friedman's *The Lexus & the Olive Tree.* Year 2 expanded the framework to include economics, geography, and government and used Friedman's *The World is Flat* to examine the interactions between the three social studies curricular areas. Year 3—summer 2007—examined economics, geography, government, and sociology and how global economics impacts Indiana. The centerpiece of Year 3 examines how the spaces of globalization alter the landscape of Indiana communities (or places) within the context of economic development, politics, and social structures. This is based on Friedman's concept of glocalization: the interaction of localities and the global economy. The primary unifying geographic strand is "Why regions matter?" and how globalization results in a discreet politics of scale that empowers places to grow or makes them vulnerable to uneven development. The workshops are designed to help teachers understand the evolution of social systems that is occurring and what communities (at the national, state, and local levels) need to do to become more successful in the current social system we call globalization, and to become familiar with techniques and resources to transmit this understanding to their students.

The basic foundation of integrating geography into the economic curriculum are two National Geographic Themes—regions and movement. To understand regions, the regional concept and types of regions are used to explore how globalizing forces are redefining formal, functional, and perceptual regions. The primary forces include the movement of goods, services, and data from one region to another and the ability of new technology to transcend distance and eliminate previously existing barriers to spatial interaction. The discussion of why regions matter examines how mobile labor, international trade, and telecommunications infrastructures have transformed the world economy—as well as global governance structures (such as free trade) and local cultures (consumerism). To demonstrate these themes, the integration of geography focuses on regional trading partnerships and economic development.

Regional trading partnerships are examined as functional regions. Prominent regions, such as the European Union (EU)—are discussed using a variety of education resources created by the EU. In addition to discussions of region types, the integration of geography into economics can also benefit from a discussion of the space-place, or local-global dialectic within the context of uneven development. For example, Map 10.5 is much more than a cartographic representation of the EU. Beyond dates of entry, the chloropleth map intuitively refers to differential degrees of economic

Map 10.5. European Union. *Source:* European Union (2002) Retrieved from http://europa.com.

development and the degree of social, political, and economic transformation required to enter the EU. In the case of those states colored "green" (Turkey, Bulgaria, Hungary, and states of the former Yugoslavia), the degree of transformation required to integrate into the European economy was greater than the core states in Western Europe.

To further understand the geography of world trade, the World Trade Organization (WTO) has created a series of maps that charts the regions of the global world. Interestingly, the WTO maps do not necessarily correspond to either a map of the continents—but tend to correspond to a combination of trade and political considerations (Map 10.6). In the case of Europe, the WTO geography mirrors the geography of the EU. In contrast, North America does not correspond to the geography of the North American Free Trade Agreement. As such, the regional geographies that have emerged as a result of globalization reflect some of the uncertainties of era—as well as the historical context of individual nations and regions. For example, why has Mexico been excluded from North America—when the EU defines the WTOs Europe? Likewise, why has the WTO created (or perpetuated) a division between Western and Eastern Europe? These basic geographic questions can be used by social studies teachers to unpack a wide array of concepts from across the social sciences. In the process of answering these questions, teachers will be able to address the full array of content standards including psychology, sociology, geography, government/civics, history, and economics.

In the end, geography—as an integrative curricular framework—has the potential to enrich the existing curriculum and prepare students for the

Map 10.6. WTO regions. *Source:* World Trade Organization, 2003 (http://www.wto.org, accessed 15 August 2006.)

local and global uncertainties of globalization. In the process, geography—when coupled with digital resources—can be used to explore the place-based realities of globalization in local communities and the hyper-competitive spaces of globalization.

CONCLUSION

The digital classroom and the availability of GST have the potential to provide students with rich learning experiences. However, the use of these technologies within the classroom should be within a specific conceptual framework that is standards-based, disciplinarily engaged and socially relevant. As our two examples of integrating geography across the curriculum demonstrate, a variety of digital resources can be used to teach the geography standards. Indeed, the use of GST in the classroom should be conceptually driven and standards based—not technology oriented per se (Gatrell, 2001, 2004). As such, the range of resources available to geography, economics, and history teachers is much broader than simply GIS, remote sensing, and GPS. In the end, the digital classroom has the potential to change the role of geography within the social studies curriculum through the infusion of geographic principles across multiple courses.

ACKNOWLEDGEMENTS

The work described in this paper has been supported by grants from the National Council for Economic Education through funding from the US Department of Education Office of Innovation and Improvement and the Indiana Department of Education. The opinions expressed in this paper are solely the responsibility of the authors.

NOTE

1. The Indiana Core40 is the standard high school curriculum in the state and was developed collaboratively with members of the Education Roundtable. Members include leaders from business, labor, education, and elected officials. The roundtable was created as a response to the changing dynamics of the global economy and has come to embody many of the curricular reforms outlined in the "Rising above the Gathering Storm" report.

REFERENCES

Engels, F. (1892). *The condition of the working-class in England in 1844.* London: Swan Sonnenschein.

Friedman, T. (1999). *The lexis and the olive tree.* New York: Farrar, Straus, and Giroux

Friedman, T. (2005). *The world is flat: A brief history of the 21st century.* New York: Farrar, Straus, and Giroux

Gatrell, J. (2001). Structural, technical, and definitional issues: The case of geography & GIS in the K–12 classroom. *The Journal of Educational Technology Systems, 29*(3), 237–249.

Gatrell, J. (2004). Making room: Integrating geo-technologies into teacher education. *Journal of Geography, 103,* 193–198.

Gatrell, J., & Collins-Kreiner, N. (2006). Negotiated space: Tourists, pilgrims and the Bahai Gardens in Haifa. *Geoforum, 37,* 765–778.

Lefebvre, H. (1991). *The production of space.* Oxford: Blackwell.

McCann, E. (1999). Race, protest, and public space: Contextualizing Lefebvre in the US city. *Antipode, 31,* 163–184.

Merrifield, A. (1993a). The struggle over place: Redeveloping American can in Baltimore. *Transactions of the Institute of British Geographers, 18,* 102–121.

Merrifield, A. (1993b). Place and space: A Lefebvrian reconciliation. *Transactions of the Institute of British Geographers, 18,* 516–531.

National Academies of Science, National Academy of Engineering, and Institute of Medicine (COSEUP). (2005). *Rising above the gathering storm: Energizing and employing America for a brighter future.* Washington, DC: National Academies.

Warf, B. (1999). Constructing a dialogue: Geographic education and the geographic research. *The Professional Geographer, 51,* 586–591.

CHAPTER 11

COMING OF AGE

Research and Pedagogy on Geospatial Technologies within K–12 Social Studies Education

Aaron Doering
University of Minnesota

George Veletsianos
University of Minnesota

Cassandra Scharber
University of Minnesota

Throughout the past century, technology use within education has gone through numerous iterations with each new technology promising a transformation for learners, instructors, and classrooms. From the audiovisual movement in the 1930s, the computer-assisted instruction movement in the 1970s and 1980s, and the Internet era of today, the field of educational technology is continually striving to understand how to best integrate technology within educational contexts in order to enhance instruction and learning—a goal that has not been realized. The lessons that researchers within the field of educational technology have learned from the past are that no technology is a panacea for education and that teachers will always

Digital Geography: Geospatial Technologies in the Social Studies Classroom, pages 213–226
Copyright © 2008 by Information Age Publishing
All rights of reproduction in any form reserved.

be an important factor in better understanding how technology can influence learning (Clark, 1983, 1985, 1991, 1994; Roblyer & Edwards, 2000).

This trepidation about investment in and research about educational technology is also echoed within the field of social studies education. In the late 1990s, Martorella (1997) strongly urged social studies educators and their research communities to tap into the power of technology for supporting and transforming social studies teaching and learning. Although the research community has responded with a small sampling of research, the potential of technology, specifically geographic information systems (GIS), within social studies education has not been realized (e.g., Bednarz & van der Schee, 2006; Freeman, 2003). Ross (2000), editor of a special issue of technology in social studies within *Theory and Research in Social Education* noted, "until longitudinal in-depth studies incorporating technological and social studies instructional precepts are conducted, technology will only be given a brief nod of acceptance as something that is nice to think about but not a necessity within the social studies community" (p. 500). Furthermore, in response to Martorella's (1997) comment that technology within the social studies is a "sleeping giant," Bolick (2004) argued that the "giant is waking" in certain areas within the social studies and called for collaboration between social studies teacher educators, instructional designers, and technology specialists in order to realize the potential of technology within social studies education. We agree with Bolick that this type of cross-disciplinary collaboration is necessary if the affordances inherent within and through the use of geospatial technologies (GST) in the field of social studies education are to be realized. Specifically, GST have great potential to enhance the teaching and learning of geography. We believe this vision of a fully awake and functioning "giant" can be a reality through taking to heart and understanding the lessons learned about technology integration from the field of learning technologies and from using theoretical frameworks focused on pedagogy as guides for future social studies educational research.

In this chapter, we focus our discussion on technology research within the social studies by highlighting the lessons learned from the field of learning technologies, discussing the disconnect between theory and practice, discussing the current research on GST integration within the social studies, and finally, suggesting frameworks for future research on GST integration in K–12 contexts.

LESSONS FROM LEARNING TECHNOLOGIES

We fully recognize that most often research does not tend to cross disciplines. However, the field of learning technologies can inform the field of social studies education. Through the lessons learned in the field of learning technologies, we believe that learning technologies research on pedagogy and technology can inform social studies educational research.

Within learning technologies, educators and researchers tend to hope for widespread transformations in education because technological advancements have brought about such pervasive impact in numerous other areas of society in the past (Bull, Knezek, Roblyer, Schrum, & Thompson, 2005). For example, in 1922 Thomas Edison proclaimed that the "motion picture is destined to revolutionize our education system and that in a few years it will supplant largely, if not entirely, the use of textbooks" (Brooker, 1947, p. 103); Seymour Papert (1984) asserted the computer as "a catalyst of deep and radical change in the educational system"(p. 422); and Cisco Systems Inc. CEO John Chambers noted that "education over the Internet is going to be so big it is going to make e-mail usage look like a rounding error" (Friedman, 1999, p. 25). The expectation that technological innovations will and should redefine the educational landscape is still rampant (e.g. Federation of American Scientists, 2006), even though history has shown that these grandiose visions about technology in education have not come to fruition (Cuban, 2001). Additionally, the convincing historical evidence presented by Russell (2001) further illustrates that technology has not transformed teaching and learning. Since 1928, 355 research studies found "no significant differences" in media comparison studies. Media comparison studies are those studies that compare student outcomes across two media (e.g. face to face and distance delivery courses). In other words, these studies found that the two media being compared are no different in terms of student outcomes.

The media debate (Clark, 1983, 1994), albeit dated, is alive and well today as evidenced by the content of journal articles throughout learning technologies and content-specific subjects such as social studies (Robinson & Nathan, 2001). On the one hand, some educational researchers are referencing the debate to support why their research is moving away from media-comparison studies and shifting to focusing on the affordances of effective pedagogy with technology in the classroom (Cronjé, 2006; Doering, 2006). On the other hand, within social studies education, some researchers continue to conduct classroom technology research using cross-media approaches. If we continue to study GST in this manner, neither the pedagogical nor the technological affordances of GST will become apparent. As a result, GST may not be utilized effectively, thus inhibiting its full potential within the classroom. *We must therefore shift the focus of social studies educational research to investigate the most effective pedagogies using GST (e.g. GIS) in the ways it was developed—as a constructivist tool.*

DISCONNECTIONS: THEORY, GEOSPATIAL TECHNOLOGY AFFORDANCES, AND PRACTICE

We view GST as "constructivist technologies" or "mindtools" (Jonassen, 1995, 2000). Inherently, GST do not possess any content value, but the value comes from the student taking the role of a geographer and the teacher taking the role of an instructional designer (Dexter, 2002). GST were designed for agency and industry use, therefore, students learn not *from* GST, but *with* GST (Jonassen, 2000).

Unlike traditional uses of instructional software that is produced and then delivered *to* students, GST enable students to be in the designer's seat as they construct meaning. In this vein, we must remember that constructivism is a theory of learning, not a theory of teaching. Thus, both education and research utilizing GST should center on students constructing their own knowledge and participating in the solving of problems based on authentic contexts. This constructivist approach to learning within educational technology has become very prominent in the literature within the field of learning technologies at the college level (Duffy & Cunningham, 1996), but questions remain about what is actually transpiring within K–12 classrooms. Is there a disconnect between the inherent design of GST, the theoretical approach encouraged within the classroom, and the actual use of these technologies within the classroom?

In general, many educators would argue that current educational practice does not follow from theory as a result of issues such as standards testing, the investment of time, the individual learning paths of students, and the difficulty of assessment when each student is truly constructing his/her own knowledge (Polin, 2004). Although the use of GST has been noted to assist learners in meeting *all* of the National Geography Standards (Audet & Paris, 1997; Bednarz, 1995) and integration of them in K–12 schools is growing, integration is still not at a rate that educators and software developers had once hoped (C. Fitzpatrick, personal communication, 2006). In addition, state integration of geography and GIS are at best uneven and under-represented across the U.S.

Numerous reasons have been noted within the literature for this disconnect and the continued slow integration within the K–12 classroom including the lack of a research agenda in K–16 GIS education, a lack of quality research, and need for more trained researchers (Baker & Bednarz, 2003). Charlie Fitzpatrick, Co-Manager of K–12 Education for the Environmental Systems Research Institute, the company that develops GST such as ArcView and ArcExplorer Java Edition for Education, commented, "[the integration of GIS] is behind where we would like to see it. We know there are people who are doing a great deal with it and it isn't necessarily that the technology as a whole is behind, but the implementation of it is behind."

(Personal communication, 2006). Fitzpatrick also discussed the need for models of pedagogy and social contexts that support the technology. He further commented, "some people are able to do more powerful things. They are in a situation where the administration supports the use of it. There are a number of people who are following a strict set of lessons and others who are encouraging students to go off exploring. Some teachers are going to do what is right and those that are going to do what they are told." This disconnect between the affordances of the technology and its slow integration into K–12 classrooms is common throughout schooling when it comes to constructivist technologies. Teachers require both freedom and support to create learning environments within their classrooms that support constructivist learning with GST. This can be a tall order in today's K–12 schools.

How do we get to the point where teachers are teaching and students are learning in this fashion? We do so through advancing the research about what works and what does not work pedagogically when using GST in the K–12 classroom. In other words, an articulation of training and research designs are needed.

RESEARCH "LIGHT" IN THE SOCIAL STUDIES

In an overview of the research in social studies education regarding the use of technology, Friedman and Hicks (2006) comment that the field can be "criticized as being 'research light'" (p. 251). In order to shed this criticism, geospatial, K–12 social studies research should be situated within positivistic, interpretive, and critical paradigms so that findings can, 1) be generalized across populations (if possible), and 2) provide rich descriptions of pedagogical practices and methodologies so practitioners and researchers can inform practice. To obtain a more holistic perspective of past and current research on GST within social studies education, we searched digital databases, (i.e. Education Full Text, Education Resources Information Center (ERIC), and Digital Dissertations), limiting our search to peer-reviewed articles published between 2000 and 2006. We also examined various handbook chapters and conference proceedings from technology and social studies conferences. Additionally, leading research and practitioner journals in social studies education were consulted including *Theory and Research in Social Education* and *Social Education*.

On the whole, the research on GST within social studies education has been described as "adolescent" (Berson & Balyta, 2004). The research appears thin and exploratory in nature; articles and papers typically identify no methodological or theoretical frameworks; offer rationalizations for the use of GST; introduce and describe online geospatial software;

make recommendations on the use of GST and proposals for integrating these into the curriculum; and propose uses and benefits of geospatial software in research and the classroom. Our examination of the literature also indicates a general lack of concern for rigorous empirical examination of how GST may impact learning. As noted by Baker and Bednarz (2003), research on GIS in education needs to involve a research agenda, with suitable research questions, designs, data, and methodologies. We emphasize the need for rigorous research in this area because we observed methodological and research design limitations in the published manuscripts that we reviewed.

It should also be noted that a number of authors recognized and accounted for the importance of pedagogy in their manuscripts. For example, Baker and White (2003) and Ramos, Miller, and Korfmacher (2003) utilized a problem-based learning approach; Beckett and Shaffer (2005) employed a pedagogical praxis stance; and Wiegand (2003) engaged students in collaborative learning activities with GIS acting as a supportive tool. Although Baker and White (2003) compare a group of learners using a collaborative GIS and a group of learners using paper maps, they recognize that the positive effects observed for the GIS group cannot be wholly attributed to the technology. Further, we believe that the affordances provided by the media supplied to the two groups are inherently dissimilar, making comparisons between the two groups perilous. Yet, these four studies appear to use technology as a supportive tool in learning activities as proposed by Jonassen, Campbell, and Davidson (1994) by considering pedagogy and learning theory prior to implementing a GIS solution. We propose that the research on GST in K–12 classrooms needs to push through its adolescence and come of age. "Research needs to be research" (Baker & Bednarz, 2003, p. 233), regardless of epistemological paradigm (i.e. positivistic, interpretive, or critical).

FRAMEWORKS FOR FUTURE RESEARCH

Based on the lessons learned from the fields of learning technologies and social studies education, it is time to take advantage of the strengths of the "giant" within social studies education. To achieve this vision, we offer several frameworks to guide research on GST within the social studies. These frameworks are focused on pedagogical, methodological and theoretical research models. Pedagogical models are important because they can reveal the most effective ways to teach with technology in social studies. Pedagogy-centered frameworks have the potential to simultaneously spur innovations in social studies instruction and new directions in research surrounding technology use in social studies teaching and learning. Social studies and geography instruction has changed little in decades despite the additions and possibilities offered by technology (e. g. Bednarz & van der Schee, 2006; Diem, 2002; Glenn, 2002; Martorella, 1997). Therefore, as Glenn (2002) suggested, GST themselves are not the problem; it is social studies instruction that is problematic. This is a *philosophical* issue: "what must be confronted at some point are social studies educators' beliefs about knowledge, teacher and student roles, and fundamentals of learning" (Glenn, 2002, p. 181). Establishing methodological frameworks is important because they provide systematic approaches to studying a phenomenon of interest. As such, the use of commonly accepted methodologies allows for consistency across, and comparability and replicability between studies. Finally, theory informs practice and aids us in applying our ideas to the real world. We need good theory "in order to bridge the gap between description and design" (Mishra & Koehler, 2006, p. 1045).

There are multiple frameworks that can guide future research on the use of implementation of GST. We advocate the use of the following pedagogical, theoretical, and methodological frameworks to guide future social studies research: a) *models of geospatial pedagogy*; b) *geography technological pedagogical content knowledge* (G-TPCK) (Doering and Veletsianos, in press) based on Shulman's (1986) initial work on pedagogical content knowledge; c) *constructivism* to understand how students can effectively construct knowledge with technology (Jonassen, 2000); and d) *design-based research* in order to truly understand learning in context (Brown, 1992; Collins, 1992).

Geospatial Pedagogical Models

As noted by the previous research within learning technologies, research in social studies education should not focus on comparing the presence of technology versus the absence of technology, but rather on the most effective pedagogies to teach with technology, specifically GST. Little research has been done within this area for notable reasons. First, to accomplish quality research within the social studies related to GST, researchers need money. Historically, the majority of government funds have been funneled to math and science. Currently, geography is the only core academic subject recognized in the No Child Left Behind legislation that does not have dedicated federal funding. Second, to research multiple pedagogies, researchers need a large number of participants within the study to statistically compare both between and across group membership; this situation most likely will not happen without grant money to fuel the research for large-scale studies.

An example of geospatial pedagogical research is proposed by Doering (2004) who designed three pedagogical models for teaching preservice teachers how to learn and teach GIS within their future K–12 classrooms. Three pedagogical models—basics first, structured problem solving, and guided generation—were developed as part of an online learning environment to be used within a preservice teacher education program and the K–12 classroom. These models included a traditional instructionist approach (basics first), constructivist scaffolded approach (structured problem solving), and a constructivist exploratory approach (guided generation). Students learned significantly more when instructed using the structured problem solving model retaining both declarative and procedural knowledge significantly better two weeks after instruction.

The benefits of Doering's (2004) pedagogically-focused research are that it may provide teacher education programs with guidelines on how to effectively teach technology to preservice teachers and it assists pre- and inservice teachers who have not been trained with GST by providing models and strategies to teach GST within their future classrooms. However, the research should not stop here. In addition to these pedagogical models, teachers need to develop geography technological pedagogical content knowledge (Doering and Veletsianos, in press).

Geography Technological Pedagogical Content Knowledge (G-TPCK)

Although the focus of teacher education historically has been on teachers' content knowledge (Schulman, 1986), there has been a vital shift to focusing on teachers' pedagogy (Ball & McDiarmid, 1990). Schulman (1986) provided an in-depth look at what he called pedagogical content knowledge (PCK), three components of teacher knowledge, and stressed the importance of not looking at these components separately. Hughes (2000) took pedagogical content knowledge (PCK) one step further by adding technology as another component of knowledge that is needed by teachers—technological pedagogical content knowledge (TPCK). Most recently, Mishra and Koehler (2006) have discussed the TPCK theoretical framework in greater depth while Hughes and Scharber (in press) discuss TPCK within the content area of English. Doering and Veletsianos (in press) discuss the need for social studies teachers to have geography technological pedagogical knowledge (G-TPCK). In this view, geography teachers understand more than technology alone (e.g. Google Earth), more than pedagogical models alone (e.g. structured problem solving), and more than content alone (e.g. cultural geography). Rather, they understand the complex relationships and interconnections between and among all three of these components of teacher knowledge (see Figure 11.1).

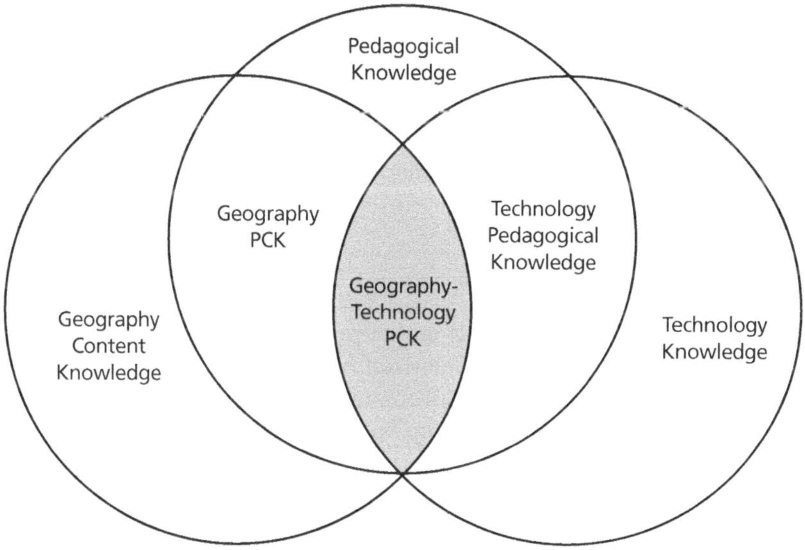

Figure 11.1. Geographical Technological Pedagogical Content Knowledge (G-TPCK)—adapted and extended from Hughes (2000).

Constructivist Learning

If GST are seen and used as constructivist tools, the result for students is constructivist learning. Although Armento (1986) identified constructivism as a promising research focus, questions remain around how we should study this area of learning where students are engaged through constructing their own knowledge with GST. Seixas (2001) stated that, "the ambivalent relationship of social studies to its component academic disciplines poses a problem for this research program. Without models drawn from the disciplines of history and the social sciences, potential frameworks for studying the development of students' ideas remain murky" (p. 546). Thus, not only do we need pedagogical models for guiding social studies learning and teaching, we need models for how to research the resulting constructivist learning.

The use of GST as constructivist tools implies that conceptions of education need to change. Simply stated, our schools and classrooms need to focus on meaningful learning with technology—learning that is grounded in problem solving where students are learning *with* technology (Jonassen, 1995, 2000). This vision of education requires teachers to relinquish control as they assume the role of facilitators assisting students in learning with the GST. For the students, this vision means they assume more authority in the classroom as they learn to articulate, reflect, and evaluate geographic problems using the technology (Alibrandi, 2003). In terms of learning with GST, this vision means leading students to ask geographic questions, acquire geographic resources, explore geographic data, analyze geographic information, and act upon the geographic knowledge they developed in concert with the technology and each other (Malone, Palmer & Voigt, 2002; National Geography Education Standards Project, 1994).

Design-Based Research

Design-based research (DBR) holds promise for research on GST in social studies. DBR can be characterized by educational interventions concerned with solving real-world problems (Wang and Hannafin, 2005). More formally, DBR is a multi-step methodological approach aimed at enhancing learning and teaching processes by means of theory development, research in authentic and naturalistic environments, and the sharing of knowledge amongst practitioners and researchers (The Design-Based Research Collective, 2003). In terms of the research process, DBR is interactive, iterative, and flexible (Wang and Hannafin, 2005) because data can be gathered before, during, and after the technological/pedagogical intervention and utilized to revise and refine both the intervention and the theory. Finally, DBR is situated and contextual because the research results are associated with the setting where the research was conducted, and multi- and cross-paradigmatic as researchers may need to utilize research methods from a variety of research paradigms.

The DBR framework is a worthwhile methodological endeavor to pursue in social studies educational research about technology use and integration because researchers need to move away from one-shot approaches to designing interventions, and instead examine ecologically valid learning and teaching processes. As emphasized previously, social studies education researchers need to focus on pedagogy and DBR appears to be well-suited for rethinking the pedagogy with which we learn and teach GST in the K–12 social studies classroom. Furthermore, by collaborating with teachers and immersing ourselves in contextual and authentic environments, educational researchers and teachers together may be able to advance theory, teaching, and learning with GST in K–12 classrooms beyond its current "adolescent" stage (Berson & Balyta, 2004).

SUMMARY

In this chapter we presented an examination of GST use and research within social studies, and proposed that focus should be directed on pedagogy rather than technology. We concluded by recommending four frameworks that we deem valuable for future GST research in social studies education.

We wrote this chapter in the hopes of assisting GST research within social studies to "come of age." We feel that focused and rigorous research (regardless of epistemological standpoint) will help advance integration of GST in the K–12 environment, provide practitioners and researchers with exemplary teaching practices, and illuminate how students learn when using GST in authentic, contextual, and ecologically-valid settings. Additionally, we hope to encourage an epistemological shift in how teachers and students think. We hold that knowledge cannot be viewed as a static artifact to be given or passed to students through osmosis. Rather, knowledge should be viewed as self-created, self-relevant, self-structured, and authentic. Given such a definition, we hope that we will provide the impetus for the awakening of the sleepy, yet mighty giant within social studies education.

REFERENCES

Alibrandi, M. (2003). *GIS in the classroom: Using geographic information systems in social studies and environmental science.* Portsmouth, NH: Heinemann.

Armento, B. (1986). Research on teaching social studies. In M.C. Wittrock (Ed.), *Handbook of research on teaching* (3rd ed.) (pp. 942–951). New York: Macmillan.

Audet, R. H., & Paris, J. (1997). GIS implementation model for schools: Assessing the critical concerns. *Journal of Geography, 96,* 293–300.

Baker, T., & Bednarz, S. (2003). Lessons learned from reviewing research in GIS education. *Journal of Geography, 102,* 231–233.

Baker, T., & White, S. (2003). The effects of GIS on students' attitudes, self-efficacy, and achievement in middle school science classrooms. *Journal of Geography, 102*(6), 243–254.

Ball, D. L., & McDiarmid, G. W. (1990). The subject matter preparation of teachers. In W.R. Houston (Ed.), *Handbook of research on teacher education* (pp. 437–449). New York: Macmillan.

Beckett, K. L., & Shaffer, D. W. (2005). Augmented by reality: The pedagogical praxis of urban planning as a pathway to ecological thinking. *Journal of Educational Computing Research, 33*(1), 31–52.

Bednarz, S. W. (1995). *Reaching new standards: GIS and K–12 geography.* Retrieved April 10, 2001, from: http://www.odyseey.maine.edu/gisweb/spatdb/gislis95/gi95006.html

Bednarz, S. W., & van der Schee, J. (2006). Europe and the United States: The implementation of geographic information systems in secondary education in two contexts. *Technology, Pedagogy, and Education, 15*(2), 191–205.

Berson, M. J., & Balyta, P. (2004). Technological thinking and practice in the social studies: Transcending the tumultuous adolescence of reform. *Journal of Computing in Teacher Education, 20*(4), 141–150.

Bolick, C. (2004). The giant is waking! *Journal of Computing in Teacher Education, 20*(4), 130–132.

Brooker, F. (1947). Motion pictures as an aid to education. *Annals of the American Academy of Political and Social Science, 254*, 103–109.

Brown, A. L. (1992). Design experiments: Theoretical and methodological challenges in creating complex interventions in classroom settings. *The Journal of the Learning Sciences, 2*(2), 141–178.

Bull, G., Knezek, G., Roblyer, M. D., Schrum, L., & Thompson, A. (2005). A proactive approach to a research agenda for educational technology. *Journal of Research on Technology in Education, 37*(3), 217–220.

Clark, R. (1985). Evidence for confounding in computer-based instruction studies: Analyzing the meta-analyses. *Educational Communications and Technology Journal, 33*(4), 249–262.

Clark, R. (1991). When researchers swim upstream: Reflections on an unpopular argument about learning from media. *Educational Technology, 31*(2), 34–40.

Clark, R. (1994). Media will never influence learning. *Educational Technology Research and Development, 42*(2), 21–29.

Clark, R. E. (1983). Reconsidering research on learning from media. *Review of Educational Research, 43*(4), 445–459.

Collins, A. (1992). Toward a design science of education. In E. Scanlon & T. O'Shea (Eds.), *New directions in educational technology* (pp. 15–22). New York: Springer-Verlag.

Cronjé, J. (2006). Paradigms regained: Toward integrating objectivism and constructivism in instructional design and the learning sciences. *Educational Technology Research & Development, 54*(4), 387–416.

Cuban, L. (2001). *Oversold and underused: Reforming schools through technology 1980–2000.* Cambridge, MA: Harvard University Press.

Design-Based Research Collective. (2003). Design-based research: An emerging paradigm for educational inquiry. *Educational Researcher, 32*(1), 5–8.

Dexter, S. (2002). eTIPS-Educational technology integration and implementation principles. In P. Rodgers (Ed.), *Designing instruction for technology-enhanced learning* (pp. 56–70). New York: Idea Group Publishing.

Diem, R. A. (2002). It was twenty years ago today: Prospects for infusing technology in the social studies. *The International Social Studies Forum, 2*(2), 171–173.

Doering A. (2004). *GIS in education: An examination of pedagogy.* Unpublished doctoral dissertation. University of Minnesota.

Doering, A. (2006). Adventure learning: Transformative hybrid online education. *Distance Education, 27*(2), 197–215.

Doering, A., & Veletsianos, G. (in press). An investigation of the use of real-time, authentic geospatial data in the K–12 classroom. *Journal of Geography,* Special issue on Using Geospatial Data in Geographic Education.

Duffy, T. M., & Cunningham, D. J. (1996). Constructivism: Implications for the design and delivery of instruction. In D. H. Jonassen (Ed.), *Handbook of research for educational communications and technology* (pp. 170–198). New York: Simon & Schuster Macmillan.

Federation of American Scientists. (2006). *Harnessing the power of video games for learning.* Retrieved November 4, 2006, from http://fas.org/gamesummit/Resources/Summit%20on%20Educational%20Games.pdf

Freeman, D. (2003). GIS in secondary education. *Teaching Geography, 28*(1), 38–41.

Friedman, A. M., & Hicks, D. (2006). The state of the field: Technology, social studies, and teacher education. *Contemporary Issues in Technology and Teacher Education* [Online serial], *6*(2). Available: http://www.citejournal.org/vol6/iss2/socialstudies/article1.cfm.

Friedman, T. L. (1999, November 17). Foreign affairs; Next, it's e-ducation. *New York Times*, pp. 25, A5.

Glenn, A. D. (2002). Thirty years and some of the same issues. *The International Social Studies Forum, 2*(2), 179–181.

Hughes, J. E. (2000). *Teaching English with technology: Exploring teacher learning and practice.* Unpublished doctoral dissertation, Michigan State University, East Lansing, MI.

Hughes, J. E., & Scharber, C. (In press). Leveraging the development of English-technology pedagogical content knowledge within the deictic nature of literacy. In *Handbook of Technological Pedagogical Content Knowledge for Teaching and Teacher Education.* Monograph by AACTE's Committee on Innovation and Technology. Lawrence Erlbaum.

Jonassen, D. (1995). Computers as cognitive tools: Learning with technology, not from technology. *Journal of Computing in Higher Education, 6*(2), 40–73.

Jonassen, D. (2000). *Computers as mindtools for schools: Engaging critical thinking.* Columbus, OH: Prentice Hall.

Jonassen, D., Campbell, J., & Davidson, M. (1994). Learning with media: Restructuring the debate. *Educational Technology, Research and Development, 42*(2), 31–39.

Malone, L., Palmer, A., & Voigt, C. (2002). *Mapping our world: GIS lessons for educators.* Redlands, California: ESRI Press.

Martorella, P. (1997). Technology and the social studies: Which way to the sleeping giant? *Theory and Research in Social Education, 25*(4), 511–514.

Mishra, P., & Koehler, M. (2006). Technological pedagogical content knowledge: A framework for teacher knowledge. *Teachers College Record, 108*(6), 1017–1054.

National Geography Education Standards Project. (1994). *Geography for life: National geography standards, 1994.* Washington, DC: National Geographic Research & Exploration.

Papert, S., (1984). New theories for new learnings. *School Psychology Review, 13*(4), 422–428.

Polin, L. (2004). Learning in dialogue with a practicing community. In T. M. Duffy & J. Kirkle (Eds.), *Learner-centered theory and practice in distance education: Cases from higher education* (pp 17–48). New Jersey: Erlbaum.

Ramos, B., Miller, S., & Korfmacher, K. (2003). Implementation of a geographic information system in the chemistry laboratory: An exercise in integrating environmental analysis and assessment. *Journal of Chemical Education, 80*(1), 50–53.

Robinson, C., & Nathan, M. (2001). Considerations of learning and learning research: Revisiting the "media effects" debate. *Journal of Interactive Learning Research, 12*(1), pp. 69–88. Norfolk, VA: AACE.

Roblyer, M. D., & Edwards, J. (2000). *Integrating educational technology into teaching* (2nd ed.). Upper Saddle River. NJ: Merrill.

Ross, E. (2000). The promise and perils of e-learning. *Theory and Research in Social Education, 28*(4), 482–492.

Russell, T. (2001). *The no significant difference phenomenon: A comparative research annotated bibliography on technology for distance education.* Montgomery, AL: IDECC.

Seixas, P. (2001). Review of research on social studies. In V. Richardson (Ed.), *Handbook of Research on Teaching, 4th ed.* (pp. 545–565). American Educational Research Association

Shulman, L. S. (1986). Those who understand: Knowledge growth in teaching. *Educational Researcher, 15*(2), 4–14.

Wiegand, P. (2003). School students' understanding of choropleth maps: Evidence from collaborative mapmaking using GIS. *Journal of Geography, 102*(6), 234–242.

Wang, F., & Hannafin, M. J. (2005). Design-based research and technology-enhanced learning environments. *Educational Technology Research and Development, 53*(4), 5–23.

CHAPTER 12

INSTRUCTIONAL GEOGRAPHIC INFORMATION SCIENCE

A Multi-Disciplinary Framework for Geospatial Technologies in Education

Rick Bunch
University of North Carolina Greensboro

Elisabeth Nelson
University of North Carolina Greensboro

Robert Earl Lloyd
University of South Carolina

Michael Kane
University of North Carolina Greensboro

Thomas Tricot
University of North Carolina Greensboro

Interest in the employment of geospatial technology as an instructional tool in K–12 social studies classrooms and other educational areas has been increasing over the past decade. A major focus has been the use of Geographic Information Systems (GIS), Global Positioning Systems (GPS), and

Digital Geography: Geospatial Technologies in the Social Studies Classroom, pages 227–245
Copyright © 2008 by Information Age Publishing
All rights of reproduction in any form reserved.

related geospatial technology to convey basic geographic concepts to students in elementary, middle, and secondary education (Audet & Abegg, 1996; Bednarz & Audet, 1999; Broda & Baxter, 2003). Classroom instruction that uses this technological potpourri to teach basic geographic concepts can be aptly called *Instructional Geographic Information Systems (InGIS).* The impetus for *InGIS* is partially rooted in the dramatic improvements and reduction of costs in technology and the introduction of educational GIS software purchase programs. Many of these programs have provided an economical and feasible platform for purchasing and incorporating GIS into K–12 classrooms throughout the United States. In the spring of 2005, the United States Department of Labor outlined a host of current and future employment opportunities in geography, with the caveat that there will be a shortage of individuals qualified to fill the needs in many geospatial technical areas (Crosby, 2005). Moreover, geospatial technology has been recently recognized as "one of the three most important and evolving fields, along with nanotechnology and biotechnology" (Gewin, 2004, pg. 376). These advances and the development of new applications for existing technology continue to increase the demand for individuals who are trained to use geospatial technology for interpreting and evaluating the world's complex spatial phenomena.

Unfortunately, successful exploitation of geospatial technology for instruction at all grade levels has been mixed and a number of substantial problems have emerged. Many teachers, for example, struggle with the complexities of learning the software or find it difficult to identify the role of GIS in the curriculum (Lloyd, 2001; Milson, DeChano, Bunch, Caito, & Qui, 2005). Teachers may also spend an enormous amount of time finding the best instructional strategies to support the use of GIS only to discontinue their development out of frustration (Kerski, 2003; Lloyd, 2001). Students who do receive lessons with geospatial technology are, in many cases, preoccupied with learning the software and the concept itself often gets lost in confusion (Audet & Paris, 1997; Cunningham, 2005; Kerski, 2003; Meyer, Butterick, Olkin, & Zack, 1999). Perhaps more importantly, there is very little agreement among educators, as well as researchers, about what age or grade level is appropriate for introducing geospatial technology, and whether the technology actually improves the learning experience (Audet & Paris, 1997; Battersby, Golledge, & Marsh, 2006; Bunch, 2000; Cunningham, 2005; Gatrell, 2004; Lloyd & Bunch, 2003). A number of solutions, ranging from pre- and in-service training for teachers to various strategies for instruction, lesson plan development, and easier-to-use GIS interfaces have been offered, but many have provided inconsistent if not poor results (Audet & Abegg, 1996; Baker & White, 2003; Blaser, Sester, & Egenhofer, 2000; Gatrell, 2004; Milson et al., 2005).

Training the trainer, instructional strategies, and lesson plan development only address part of a much bigger problem. Many instructional approaches have used geospatial technology as a stand-alone educational tool with little emphasis on the extremely important underlying geographic concepts (Gatrell, 2004). Geospatial technology is used to analyze and understand the spatial relationships among phenomena distributed on the Earth's surface. One must be able to think spatially, grasp geographic concepts, and understand the world from a spatial perspective. Effective instructional use of geospatial technology should, therefore, focus on concept instruction and learning while incorporating an understanding of how spatial information are mentally learned, processed, and retrieved by students of varying ages and spatial abilities (Albert & Golledge, 1999). This approach requires the application of theories derived from collaborative experimental studies and classroom assessments that integrate views from three relevant and major disciplines—geography, psychology, and education.

The purpose of this chapter is to provide a framework for understanding the use of geospatial technology as an instructional tool in elementary, middle, and secondary education. We take an integrative view by using cognitive and educational theories for examining how geospatial technology can be effectively used as a learning tool. We narrow the extremely large body of research from geography, psychology, and education to reflect how people think spatially within the context of maps, graphics, and other geographic displays. Our approach is intended to add the science to *InGIS* as well as shed light on the much broader and deeper issues associated with the use of geospatial technology for instruction.

The following sections provide an overview of literature related to geographic education, spatial cognition, and educational psychology within the context of developing what we refer to as *Instructional Geographic Information Science (InGIScience)*. Each section will discuss research related to using geospatial technology, maps, and graphics, for spatial learning and thinking.

INSTRUCTIONAL GEOGRAPHIC INFORMATION SCIENCE

Geography, psychology, and education are three well-established disciplines that offer much in the way of deepening our understanding of effective instructional use of geospatial technology. Each of these disciplines can be overlapped to form equally important and cross disciplinary areas of research (Figure 12.1).

Educational psychology, for example, results from the overlap of common research interests in education and psychology, and seeks to blend a

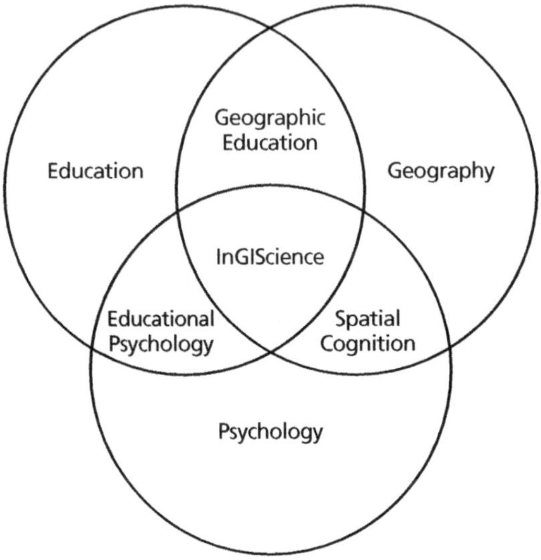

Figure 12.1. Instructional Geographic Information Science.

number of psychological and educational theories to help understand how people learn in an educational setting. Spatial cognition is the overlap of common interests between geographers and psychologists, and relies heavily on theories in geography and psychology to understand how people process, store and retrieve spatial information. Finally, geographic education—the research overlap binding geographers and educators, focuses on methodologies for improving teaching the many aspects of geography by emphasizing spatial relationships, concepts, and knowledge in an educational setting. Our proposed area of study called, *InGIScience,* is intended to provide a much richer approach by incorporating research from all three major disciplines. *InGIScience* can be further brought into focus by using research conducted in geographic education, spatial cognition, and educational psychology, all of which are intrinsically linked to three major disciplines.

Geographic Education

Geographers focus on the analysis of questions and problems from a spatial perspective, or geographic point of view. John D. Nystuen, a pioneer in spatial analysis and modern mathematical geography, defined the key concepts associated with the spatial perspective: *directional orientation, distance,* and *connectiveness* (1968). Similarly, Golledge & Stimson (1997), proposed three dimensions of spatial thinking: spatial visualization, spatial orientation, and spatial relations. The incorporation of these geographic axioms within a person's knowledge base gives them an advantage in their ability to reason in a world that is becoming increasingly connected and complex.

During the 1980s, the *Geography Education National Implementation Project* (GENIP) was established to assess the breadth and depth to which geography and its concepts were being translated to students through the nation's educational system. The National Geographic Society assembled allegiances with many other organizations to further develop a national network of interest in geography education. A poll conducted in 1988 found that Americans were far below acceptable levels in terms of geographic literacy.

Although geography is listed as a core subject within the *No Child Left Behind Act* (NCLB) of 2001, it is within the group of standards and assessments that are optional under individual state's control. In fact, geography is the only discipline without a dedicated program of provisions by the NCLB. Geography finds itself in a catch-22. NCLB measures student achievement by the use of mandatory state assessments; however, most of these assessments lack geography-related content, which mirrors the level of geography content in the classroom. Thus, the only way to be taught the material is to be tested on that material, and in order to be tested on the material, a student must be taught the material. Geography must find a way to situate itself alongside other disciplines to secure any funding. To do this, the discipline must fulfill the two top priorities for all programs: improving the academic achievement of students and providing high quality instructors. Emphasis is placed on those programs whose curricula are the product of scientifically-based research (Daley, 2003). It goes without saying that the best way to show geography's merit is to illustrate its unique and necessary role in the education of our children and to demonstrate students' effective learning of spatial skills through rigorous empirical research.

The incorporation of geographical thinking abilities into the curricula of secondary education has made modest advances over the past two decades, but there is still much work to be done if the richness of spatial analytical techniques is to help today's youth make the informed decisions needed for future success in our global society. The standards serve as educational and intellectual goals for which students should strive to achieve. Specifically, the eighteen geography standards are grouped into six foundational elements, the first of which is the ability to understand the world in spatial terms. People use maps, photographs, and satellite images to investigate the relationships between people, places, and environments. These tools display information within a spatial context. There are three geography standards that are derived from the use of these tools: (Standard 1) How to use maps and other geographic representations, tools, and technologies to acquire, process, and report information from a spatial perspective; (Standard 2) How to use mental maps to organize information about people, places, and environments in a spatial context; (Standard 3) How to analyze the spatial organization of people, places, and environments on Earth's surface (National Geography Education Standards Project, 1994).

Conspicuously underrepresented within the standards is the role of GIS in geography education. One reason for this near omission is the date of publication. In 1994, GIS was still being developed and scrutinized by the geographic community. It was initially heralded as a tool able to be utilized for the rapid and repeatable analysis inherent in geographical studies (National Geography Education Standards Project, 1994). Recently, it has gained recognition as an entity deserving consideration of becoming a major focus for its own discipline (Johnston & Sidaway, 2004). Although geographic analysis has been incorporated into the curricula of some American schools, more attention should be devoted to developing pedagogical approaches for this type of analysis and to examining the role for GIS in cultivating various forms of 'geographic thinking' among K–12 students.

GEOGRAPHIC EDUCATION AND GIS

The implementation of GIS in K–12 education has been slow due to both academic and administrative restraints. Since the inception of NCLB, teachers have been encouraged to focus classroom instruction on materials that will be tested by state standards assessments. Since geography is underrepresented on many of these tests, less time is directed to geographic concepts during classroom instruction. From an administrative perspective (excluding the initial hardware and software costs), the chief challenges most educators face in implementing GIS are a lack of training and support for the teachers, a lack of time to develop appropriate lesson plans, and the complexity of the software.

These restraints on GIS implementation and expansion are worth overcoming for several reasons. First, the incorporation of GIS may encourage students to examine data from a variety of fields. Both Jacobs (1989) and Audet & Abegg (1996) believe that an interdisciplinary approach to learning may be effective in helping students to develop problem-solving skills. The power of a GIS is that it allows us to ask questions of data. Students use an inquiry approach to form research questions, develop a methodology, gather and analyze data, and draw conclusions (National Geography Education Standards Project, 1994). Many of the models for GIS implementation in educational setting are constructivist and interdisciplinary in nature (e.g., Alibrandi, 2003; Alibrandi & Sarnoff, 2006; Baker & White, 2003; Bednarz, 2000). This allows for emphasis to be placed on the content area rather than on the system itself. Students are enabled to learn the material being presented with little distraction from the computer interface (Kerski, 2003). Palladino (1994) posited that despite its inherent spatial reasoning incentives, GIS can be utilized to enhance various teaching methods, to provide better delivery of subject matter than textbooks alone, and to prepare students for spatial technologies in the workforce.

Studies of the effectiveness of GIS for enabling the learning of spatial reasoning within students have thus far been mixed. As pointed out by Bednarz (1994) and Meyer et al. (1999), initial studies tended to focus on the ability and need for learning GIS software and not the ability of GIS to serve as a tool for enhancing students understanding of core geographic principles or axioms. Many of these studies indicate, for instance, that students develop technology skills as opposed to the spatial analytical skills sought after by the implementation of GIS. This outcome has been rationalized by many as due to the barriers of introducing GIS within the classroom—acquisition of hardware, software, and training for teachers (Meyer et al., 1999; Kerski, 2003). Studies by Meyer et al. (1999), for example, illustrate the potential setback that GIS computer software learning curve can place on students. Throughout the instruction period, considerable time was needed to learn the basic functions and capabilities of the software. Hence no difference in learning was recorded between groups of students utilizing GIS versus those who did not. Qualitative measures indicated that cartographic output was the only merit of the GIS incorporated lesson plans. In another study, however, Patterson, Reeve, & Page (2003) found that although test scores were lower than expected for all students, high school students receiving instruction through GIS scored significantly higher on test materials than college students without GIS-based instruction. Kerski (2003) also encountered mixed results when testing high school students. No significant differences were found when utilizing standardized and spatial analysis tests, although significant differences were found with respect to content-specific material.

Spatial Cognition

It is within the fields of spatial cognition and educational psychology that answers may be found for incorporating GIS into K–12 education more effectively. Spatial cognition studies have the potential to inform educators about how spatial information is most effectively acquired, processed, stored, and used. Geographers and psychologists share an interest in the cognition of space; however, they focus on the topic in different ways. Psychologists have a very broad interest in how spatial information is encoded, stored, and used (Forman & Gillette, 1997). These interests include a concern with how spatial information is acquired and represented in memory structures, how spatial abilities develop as the human brain matures, and how learning processes related to spatial cognition can be modeled (Jacobs & Schenk, 2003; Kohonen, 2001; Stiles-Davis, Kritchevsky, & Bellugi, 1988). Geographers have typically taken a narrower and more practical approach (Lloyd, 1997a). Spatial cognition studies conducted by geographers are more likely to involve real-world locations, use authentic stimuli such as maps, and give attention to practical applications.

Geographers studying spatial cognition topics have adapted many of the theoretical positions and experimental methods developed by psychologists. These adaptations provide avenues for evaluating geographic instruction and for understanding how people learn and think about space, especially from maps and geographic displays which are often generated by students and teachers who employ GIS. Studies of visual attention provide an effective starting point for examining these methodological linkages (Lloyd, 2005a; Schneider & Maasen, 1998).

Nelson (1999, 2000a, 2000b), for example, has done a series of studies that investigated how visual attention is used to understand maps and their symbols. The studies investigated how map readers obtained information from bivariate map symbols using selective attention concepts and the perceptual grouping of features. This study categorized the experimental symbols according to how the dimensions of the symbols (e.g., size and shape or hue and value) visually interact. Three types of dimensional interactions were identified with the following theoretical predictions. Map readers viewing symbols with separable dimensions should be able to attend to the two dimensions independently. Map readers cannot, however, view one integral dimension of a map symbol without processing the other integral dimension. The same map readers should be able to attend to configural dimensions of symbols independently. Configural dimensions are special, however, in that they are able to interact to produce an emergent property that takes priority over the original dimensions.

Nelson's subjects were trained to classify experimental symbols according to specific rules. When subjects applied the rules, they were required to attend to one specific dimension, selectively attend to either dimension while filtering out irrelevant dimensions, or attend to both dimensions simultaneously. Reaction times for the experimental trials were analyzed and used to design thematic maps with the aim of easing the map interpretation process in bivariate scenarios. As an example, consider a map of the spatial relationship between education and income at the county level in the United States. Symbolizing the relationship graphically with an integral symbol will enhance map reader's abilities to "see" the connections between the two datasets, while symbolizing the relationship with a symbol whose dimensions are separable will better enhance abilities to study each dataset separately.

On a related topic, additional spatial cognition studies have considered how representations of geographic space, e.g., maps and aerial photographs, are searched for information. Lloyd (1997b) constructed map symbols for his experimental map using the dimensions of color, size, shape, and orientation. Subjects sometimes searched for target map symbols that had a unique feature, compared to distracter symbols, and other times searched for target map symbols that shared features with distracter symbols. For example, a target symbol might have the unique feature red on the color dimension and other distracter symbols would have a feature other than red on the color dimension. According to cognitive search theories, target symbols with a unique feature should "pop-out" of the map, but target symbols that share features with distracter symbols should not (Duncan & Humphreys, 1992; Treisman & Gelade, 1980; Wolfe, 1994). Experimental results supported the predictions of the visual search theories, but feature differences on the color dimension produced the best "pop-out" effect, suggesting color to be one of the stronger symbol dimensions from the perspective of enhancing map interpretation.

Another visual search study that featured color as a critical variable had subjects search for color boundaries on choropleth maps (Bunch & Lloyd, 2000). Subjects were presented a pair of colors on either side of a common boundary as the target of a specific search trial. They were then presented a state map that had the counties filled with various colors that was searched to determine if the target boundary was present. Results indicated that highly luminous targets, e.g. red and yellow boundaries, were found more quickly. The similarity of the colors in the target boundary and the colors in non-target boundaries and the similarity of colors in the set of non-target boundaries also significantly affected search times.

A separate category of studies have modeled how representations of geographic space are encoded in people's mind. These types of studies provide insight into how people learn and think about space, and offer further guidance in the development of GIS as an instructional tool. We must know, for example, how students learn and process spatial information in order to teach them about spatial concepts and relationships, especially when using GIS as an instructional tool. This understanding can provide a starting point for developing GIS-based instructional strategies that address the strengths and weaknesses of students' spatial learning. Lloyd (1994), for example, investigated the learning of a prototype map, i.e. a representation of the most typical map, for a series of maps representing a particular category. Prototype maps were found to be significantly affected by the arranged sequence of individual example maps studied by subjects.

Prototypes learned by human subjects were also found to be similar to prototypes learned by auto-associator neural networks. The identification of the prototypical map for different age groups can lead to designs that incorporate this knowledge as a map template in GIS.

Other studies have used Kohonen's self-organizing map (SOM) neural network models to simulate the learning of point distributions, i.e., cities in a geographic space (Lloyd, 2000). Lloyd (2005b) had SOM neural network models learn the locations of cities in a state that were known by human subjects from three different home locations in the state. Models that had previously learned the state boundary and Interstate highway system before learning the cities consistently produced results that more closely matched the locations of the cities on sketch maps produced by human subjects.

A few studies have considered the cognitive processes related to visually processing the spatial information being presented as a product of a geographic information system. Bunch (2000), for instance, investigated how humans integrate information related to city locations mapped in a Geographic Information System. Subjects learned the information using four different experimental procedures (chunk, layer, scale, and whole) that simulated functions employed in a GIS. These procedures required subjects to integrate the spatial information across spaces, hierarchies, and geographic scales. Results indicated that the age (young adolescent or adult) and the type of integration process used to learn the information significantly affected the cognitive map of city locations learned by the subjects. This further suggests that not only the design of the geographic product being used, but also the instructional method by which it is presented and the age of the product users, may have a significant effect on the learning of spatial information.

Lloyd & Bunch (2003) used the same maps and learning procedures as Bunch (2000), but had subjects respond to True/False statements regarding the nature of the spatial information they had learned with the aid of a GIS. The reaction time, accuracy, and confidence of subjects were predicted with a back-propagation neural network model using characteristics of the learners, experimental conditions related to GIS functions, and map features (points, lines, and areas) as inputs to the model. Subjects generally were more accurate and confident in tasks that required less integration of geographical information. Young adolescent learners were slower, less accurate, and more confident than adult learners for all experimental conditions.

Albert & Golledge (1999) used an experimental design to study the cognitive abilities of subjects responding to questions related to logical operators, i.e., AND, OR, XOR, and NOT, that might be used in a GIS. Subjects worked with three related sources of information, i.e., input layer, output layer, or logical operator. They were given two of the three and required to select the correct corresponding third. Results indicated significant differences among the logical operators, but not differences based on the sex of the subjects or their prior experience with geographic information systems.

In a related study, Battersby et al. (2006) used pencil and paper tests to consider incidental learning of overlay concepts. In one study, they provided three levels of subjects (middle school, high school, and university undergraduates) with two maps (crop and soil) of the same farm and asked subjects how they would determine which areas on the farm have sandy soil and grow wheat. They were then asked to color in this area on the crop map. Since none of the subjects had any formal instruction on how to solve this problem, the study was designed to assess if subjects had incidentally learned the AND overlay process. Results indicated that university and high school subjects both used an overlay process more frequently and solved the problem more accurately. A second and related study showed the same three levels of subjects examples of AND, OR, and NOT overlay operations (Battersby et al., 2006). University and high school subjects again were generally able to solve map overlay problems better than middle school subjects. The authors concluded that the concept of map overlay is too complex to be learned incidentally by middle school students, and would therefore require instructional intervention.

Educational Psychology

It is in the educational psychology arena that we find the methodologies to evaluate the instructional methods of geographic education as paired with the design methodologies and guidelines that stem from spatial cognition. Paas & colleagues (2003b) offer *Cognitive Load Theory* (CLT) as a foundation for examining the linkages between instructional materials/ products and our abilities to use them effectively. CLT has its basis in memory, of which two types are important: limited short-term and unlimited long-term. It is the interaction between these two types of memories that is crucial, and the interactions are influenced by both task demands and spatial abilities (Paas, Renkl, & Sweller, 2003).

From a cartographic perspective, this is not necessarily a new idea; map makers have always intuitively understood the need to consider memory capacities in the context of map design. This is, in fact, the basis for several conventional design guidelines, including the simplification of information through categorization processes. What is new is the ability, through CLT, to assess the cognitive load for various map designs and instructional scenarios and to adjust designs and instructions accordingly for groups, and perhaps, individuals. Although not yet tested in *InGIScience*, CLT has worked well in several other disciplines, including physics, mathematics, computer programming and electrical engineering (Paas & Van Gog, 2006).

Cognitive Load

Cognitive load, according to Paas & van Meriënboer (1994), is influenced by both causal and assessment factors (Figure 12.2).

Individual differences, as well as task and environment are considered causal factors. Take task and environment, for example. The decision of where to best locate a new airport involves considering several criteria, from surrounding land use, to soil quality, to any other of a variety of components. This decision, however complicated, may become exponentially more so if levels of uncertainty must also be factored in for each criteria. Differences arising from changes in available geographic knowledge, then, can significantly affect cognitive load for a task.

Individual differences may also play a factor in manipulating cognitive load. Although gender is one important variable that has received attention (Hardwick. Bean, Alexander, & Shelley, 2000), studies investigating gender differences have produced mixed results (Engle & Kane, 2004). More precise differences in individual spatial abilities have been noted in theories related to brain organization (Annett, 1985, 2002), working memory capacity (Kane, Hambrick, Tuholski, Wilhelm, Payne, & Engle, 2004), prior knowledge resulting from domain experiences, or the interaction among several such factors (Casey, 1996).

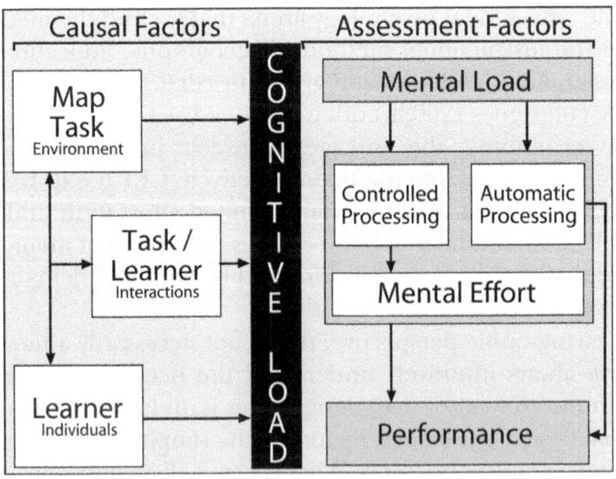

Figure 12.2. Causal and assessment factors of cognitive load (Adapted from Paas & van Merienboer, 1994).

Also affecting the cognitive load associated with tasks are assessment factors. Mental load, mental effort, and performance are measurable, and form the basis for measuring cognitive load in CLT. *Mental load* is part of the total cognitive load, and reflects the demands of task and environment in the decision-making scenario. *Mental effort* is the cognitive capacity that the learner allocates to the given task, and *performance* is the end result that reflects the *mental load*, *mental effort*, and the *causal* factors combined in a subject's decision.

Components of Cognitive Load. In addition to varying factors influencing cognitive load, it is also important to realize that recent research subdivides cognitive load into three components, each unique in what it brings to the evaluation of a cognitive task (Paas et al., 2003a). *Intrinsic* cognitive load, for example, addresses the working memory demand associated with interacting with materials and instructions. It is well-known that working memory is limited and can handle only a few interacting elements (Cowan, 2001), so this is a limiting component of cognitive load. As an example, consider the difference between learning how to apply a buffer to a linear feature and learning how to interpret the interaction of that buffer with other geographic layers to make a decision about habitat suitability. Learning to apply a buffer to a river is a low-interactivity task. Learning how to interpret the river's buffer, however, can be a high-element activity if students were required to relate the results to vegetation, human development, and other geographic influences to identify the best habitat for an endangered species.

Working to offset the limitations of working memory is long-term memory (LTM). LTM helps by allowing us to store *schemata*, which are "...cognitive constructs that incorporate multiple elements of information into a single element with a specific function" (Paas et al., 2003a, p. 2). Prior knowledge is stored in LTM and can be categorized by use (Chi, Glaser, & Rees, 1982). Prior knowledge may be stored as schemata and may help to reduce cognitive load as it can be referenced and retrieved as a single unit in memory (Schwartz, Ellsworth, Graham, & Knight, 1998). Knowing the names and locations of all the countries in the world, for example, would aid in learning the locations of all world major cities. Moreover, knowing that higher probabilities for human lead exposure exists in subdivisions built prior to 1978 would aid an analysis on blood-lead level risks in GIS.

Working to the detriment of overall cognitive load is the *extraneous* component of cognitive load. Extraneous cognitive load addresses the manner in which information is presented and the resulting learning activities required. Ambiguous instructions, for example, increase cognitive load by requiring users to search unnecessarily for the information needed to complete a task. Ambiguous classification of data on a map, poor symbology, or irrelevant data layers in a GIS are further examples.

While *extraneous* cognitive load interferes with map use, the third of the three components, *germane* cognitive load, can actually enhance learning (Paas et al., 2003a). Germane cognitive load can reference the user's motivation and/or the creation and use of schemata in performing tasks. A thorough knowledge of how GIS works and marketing can enhance *germane load*, and thus learning if an individual is exploring where to locate a new retail bank. These three forms of cognitive load are additive and together produce a total cognitive load that should not exceed the limits of working memory if learning is to be efficient and successful.

A Methodology for Measuring Cognitive Load. Cognitive load and its associated concepts must be measurable to make them useful in designing, implementing, and evaluating the instructional materials and GIS interfaces needed to promote the use of *InGIScience* to enhance geographic learning in education. One method for accomplishing this is offered by Paas and colleagues (2003a). Their method harnesses both objective and subjective measurements that are converted to z-scores and analyzed visually and statistically. Subjective measurements are typically ratings of difficulty; objective measurements are response times of users as they deal with a secondary task that is unrelated and runs concurrently with the primary task.

If a GIS user, for example is presented with a primary task of finding the intersection of a river buffer with a certain species of vegetation, they might also be asked to concurrently detect a change in the color of a letter located on the computer screen. Responses to the color change would produce reaction times indicative of how quickly the users were able to disengage from the primary task. The longer the reaction times, the higher the cognitive load of the primary task (Posner & Boies, 1971).

Objective and subjective measurements are standardized to produce a set of z-scores for each user. These are then averaged to arrive at sample means, which can be plotted on a graph. The horizontal axis of the graph represents perceived mental effort; the vertical axis represents reaction-time performance. The diagonal is a balance between the two (Paas et al., 2003a). As Figure 12.3 shows, any pair of z-score means for a task falling into the second quadrant (top left) is considered representative of a highly efficient task; here, performance exceeds effort. In contrast, a pair for a task falling into fourth quadrant (bottom right) is considered representative of low efficiency; here, effort exceeds performance. High efficiency is considered to have low cognitive load while low efficiency is considered to have high cognitive load (Brunken, Plass, & Leutner, 2003).

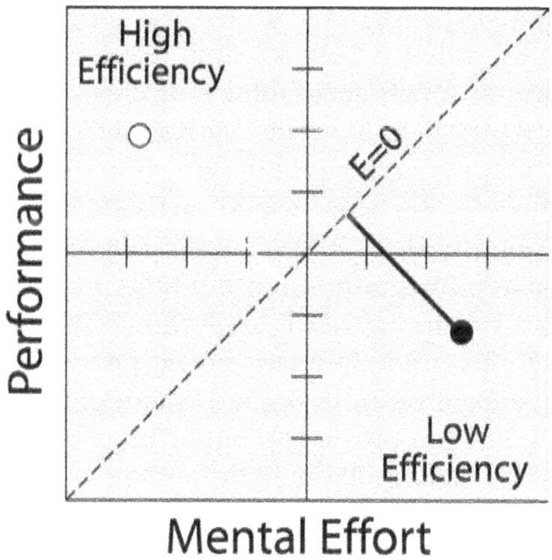

Figure 12.3. Cartesian Graphic used to visualize instructional efficiency. Group z-score means are plotted against the diagonal line (E = 0) (After Paas et al. 2003a).

CONCLUSION

Geospatial technologies are dynamic tools for studying spatial relationships between and among people, places, and objects throughout the world. This dynamic quality lends itself to the types of inquiry-based learning that is at the forefront of constructivist theory in education today. The creation, production, and analysis of interactive maps enable students to hone spatial analytical skills. The advent of the Computer Age and the Internet is allowing for the almost direct contact between cultures never before having been introduced to each other, let alone engaged in daily interaction. This phenomenon has necessitated the advancement of geographic thought in our society. This advancement must first and foremost take place in the education of our youth. Existing research from geographic education, spatial cognition, and educational psychology, taken together, can provide a formal structure for using geospatial technology as an effective educational tool. The successful citizens of tomorrow's global society will be those with the greatest understanding of the spatial interconnectedness of the world's regions and its peoples.

Previous approaches to understanding *InGIS* have been narrow and fragmented, and, as a result, its effectiveness in teaching and learning are unclear (Audet and Paris, 1997; Kerski, 2003). Perhaps exacerbating this ambiguity is the complexity of GIS itself and notion that many *InGIS* endeavors have only focused on part of the complex process e.g., the focus on application at the expense of the underlying geographic concepts. This view is supported by evidence from research that has argued for a need to understand the age at which geographic concepts can be learned before they are applied (Battersby et al., 2006). As outlined earlier in this chapter, a plethora of previous research has both directly and indirectly echoed this sentiment, and the need for more robust methods for examining *InGIS* is becoming increasingly evident. As a natural extension to *InGIS,* therefore, we argue for a richer, integrative, and multidisciplinary approach called *InGIScience.* As part of this approach, educational psychology and related cognitive load theories can help measure and provide context for understanding how students of various ages and abilities learn geographic information. This work can also provide an opportunity to identify deficiencies in spatial abilities that can be alleviated through practice, and thus increase the likelihood of effective *InGIS.* Moreover, these theories can be bolstered by research in spatial cognition that has examined the processes involved in using and learning from maps and geographic displays generated by geospatial technology. Together, these two areas of research can be placed within the context of research conducted in geographic education to focus on understanding the interaction between the instructor and students through the technology so that learning outcomes can be maximized.

Research threads found within geographic education, spatial cognition, and educational psychology can be woven together to achieve optimum educational goals. *InGIScience* has the potential to provide the overarching and integrative context for understanding the fabric of this collaborative environment because it focuses on multidisciplinary approaches. Experimentally based research grounded in *InGIScience* should be a high priority for those who wish to use geospatial technology as an educational tool.

REFERENCES

Albert, W., & Golledge, R. (1999). The use of spatial cognitive abilities in geographic information systems: The map overlay operation. *Transactions in GIS, 3,* 7–21.

Alibrandi, M. (2003). *GIS in the classroom: Using geographic information systems in social studies and environmental science.* Portsmouth, NH: Heinemann.

Alibrandi, M., & Sarnoff H. (2006). Using GIS to answer the 'whys' of 'where' in social studies. *Social Education, 70,* 138–143.

Annett, M. (1985). *Left, right, hand and brain: The right shift theory.* London: Erlbaum.

Annett, M. (2002). *Handedness and brain asymmetry: The right shift theory.* Hove, UK: Psychology Press.

Audet, R., & Abegg, G. (1996). Geographic information systems: Implications for problem solving. *Journal of Geography, 33,* 21–45.

Audet, R., & Paris, J. (1997). GIS implementation model for schools: Assessing the critical concerns. *Journal of Geography, 96,* 293–300.

Baker, T., & White, S. (2003). The effects of G.I.S. on students' attitudes, self-efficacy, and achievements in middle school science classrooms. *Journal of Geography, 102,* 243–251.

Battersby, S., Golledge, R., & Marsh, M. (2006). Incidental learning of geospatial concepts across grade levels: Map overlay. *Journal of Geography, 105,* 139–146.

Bednarz, S., & Audet, R. (1999). The status of GIS technology in teacher preparation programs. *Journal of Geography, 98,* 60–67.

Bednarz, S. W. (2000). Connecting GIS and problem based learning. In R. Audet & G. Ludwig (Eds.), *GIS in Schools* (pp. 89–102). Redlands, CA: ESRI.

Blaser, A., Sester, M., & Egenhofer, M. (2000). Visualizing in an early stage of the problem-solving process in GIS. *Computers and Geosciences, 26,* 57–66.

Broda, H., & Baxter, R. (2003). Using GIS and GPS technology as an instructional tool. *The Social Studies, 94,* 158–160.

Brunken, R., Plass, J., & Leutner, D. (2003). Direct measurement of cognitive load in multimedia learning. *Educational Psychologist, 38,* 53–61.

Bunch, R. (2000). GIS and the acquisition of spatial information: Differences among adults and young adolescents. *Research in Geographic Education, 2,* 65–95.

Bunch, R., & Lloyd R. (2000). The search for boundaries on choropleth maps: Color processing and map pattern effects. *Cartography and GIS, 27,* 15:30.

Casey, M. (1996). Understanding individual differences in spatial ability within females: A nature/nurture interaction framework. *Developmental Review, 16,* 241–60.

Chi, M., Glaser, R., & Rees, E. (1982). Expertise in problem solving. In R. Sternberg (Ed.), *Advances in the psychology of human intelligence* (pp. 7–75). Hillsdale, NJ: Erlbaum.

Cowan, N. (2001). The magical number 4 in short-term memory: A reconsideration of mental storage capacity. *Behavioral and Brain Science, 24,* 87–185.

Crosby, O. (2005). Geography jobs. *Occupational Outlook Quarterly.* The United States Department of Labor.

Cunningham, M. (2005). Why geography still needs pen and ink cartography. *Journal of Geography, 104,* 119–126.

Daley, R. (2003). *No geographer left behind: A policy guide to geography education and the no child left behind act of 2001.* Department of Geography, Texas A&M University, College Station, TX.

Duncan, J. & Humphreys, G. (1992). Beyond the search surface: Visual search and attentional engagement. *Journal of Experimental Psychology: Human Perception and Performance, 18,* 578–588.

Engle, R., & Kane, M. (2004). Executive attention, working memory capacity, and a two-factor theory of cognitive control. In B. Ross (Ed.), *The psychology of learning and motivation* (pp. 145–199). New York: Academic Press.

Forman, N., & Gillette, R. (1997). *Handbook of spatial research and methodologies.* Hove, UK: Psychology Press.

Gatrell, J. (2004). Making room: Integrating geo-technologies into teacher education. *Journal of Geography, 103,* 193–198.

Gewin, V. (2004). Mapping opportunities. *Nature, 427,* 376–377.

Golledge, R. (1999). *Wayfinding behavior: Cognitive mapping and other spatial processes.* Baltimore: Johns Hopkins University Press.

Golledge, R., & Stimson, R. (1997). *Spatial behavior: A geographic perspective.* New York: Guilford Press.

Hardwick, S., Bean, L., Alexander, K., & Shelley, F. (2000). Gender vs. sex differences: Factors affecting performance in geographic education. *Journal of Geography, 99,* 238–44.

Jacobs, H. (1989). *Interdisciplinary curriculum: Design and implementation.* Alexandria, VA: Association for Supervision and Curriculum Development.

Jacobs, L., & Schenk, F. (2003) .Unpacking the cognitive map: The parallel map theory of hippocampal function. *Psychological Review, 130,* 285–315.

Johnston, R., & Sidaway, J. (2004). *Geography and Geographers: Anglo-American Human Geography since 1945* (6th ed.). New York: Oxford University Press.

Kane, M. J., Hambrick, D. Z., Tuholski, S. W., Wilhelm, O., Payne, T. W., & Engle, R. W. (2004). The generality of working memory capacity: A latent-variable approach to verbal and visuospatial memory span and reasoning. *Journal of Experimental Psychology: General, 133,* 189–217.

Kerski, J. (2003). The implementation and effectiveness of implementing geographic information systems technology and methods in secondary education. *Journal of Geography, 102,* 128–137.

Kohonen, T. (2001). *Self-organizing maps.* New York: Springer-Verlag.

Lloyd, R. (1994). Learning spatial prototypes. *Annals of the Association of American Geographers, 84,* 418–40.

Lloyd, R. (1997a). *Spatial cognition: Geographic environments.* Dordrecht: Kluwer Academic Publishers.

Lloyd, R. (1997b). Visual search processes used in map reading. *Cartographica, 34,* 11–31.

Lloyd R. (2000). Self-organized cognitive maps. *The Professional Geographer, 52,* 517–530.

Lloyd, R. (2005a). Attention on maps. *Cartographic Perspectives, 52,* 28–57.

Lloyd, R. (2005b). Assessment of simulated cognitive maps: The influence of prior knowledge from cartographic maps. *Cartography and Geographic Information Science, 32,* 161–179.

Lloyd, R., & Bunch, R. (2003). Technology and map-learning: Users, methods, and symbols. *Annals of the Association of American Geographers, 93,* 828–850.

Lloyd, W. (2001). Integrating GIS into the undergraduate learning environment. *Journal of Geography, 100,* 158–163.

Meyer J., Butterick, J., Olkin, M., & Zack, G. (1999). GIS in the K–12 curriculum: A cautionary note. *The Professional Geographer, 51,* 571–578.

Milson, A., DeChano, L., Bunch, R., Caito, J., & Qui, X. (2005). GIS in K–12 education: Pedagogical tool or weapon of mass distraction? *Research in Geographic Education, 7,* 62–73.

National Geography Education Standards Project. (1994). *Geography for life: National geography standards, 1994.* Washington, DC: National Geographic Research & Exploration.

Nelson, E. (1999). Using selective attention theory to design bivariate point symbols. *Cartographic Perspectives, 32,* 6–28.

Nelson, E. (2000a). The impact of bivariate symbol design on task performance in a map setting. *Cartographica, 37,* 61–78.

Nelson, E. (2000b). Designing effective bivariate point symbols: The influence of perceptual grouping processes. *Cartography and Geographic Information Science, 27,* 261–278.

Nysteun, J. (1968) Identification of some fundamental spatial concepts. In B. Berry & D. Marble (Eds.), *Spatial Analysis* (pp. 35–41). Englewood Cliffs, NJ: Prentice-Hall.

Paas, F., Renkl, A., & Sweller, J. (2003a). Cognitive load theory and instructional design: Recent developments. *Educational Psychologist, 38,* 1–4.

Paas, F., Tuovinen, J., Tabbers, H., & Van Gerven, P. (2003b). Cognitive load measurement as a means to advance cognitive load theory. *Educational Psychologist, 38,* 63–71.

Paas, F., & Van Gog, T. (2006). Optimizing worked example instruction: Different ways to increase germane cognitive load. *Learning and Instruction, 16,* 87–91.

Paas, F., & Van Merrienboer, J. (1994). Instructional control of cognitive load in the training of complex cognitive tasks. *Educational Psychology Review, 6,* 351–371.

Palladino, S. (1994). A role for geographic information systems in the secondary schools: An assessment of the current status and future possibilities. Unpublished master's thesis. University of California at Santa Barbara, Geography.

Patterson, M., Reeve, K., & Page, D. (2003). Integrating geographic information systems into the secondary curricula. *Journal of Geography, 102,* 275–281.

Posner, M. I., & Boies, S. J. (1971). Components of attention. *Psychological Review, 78,* 391–408.

Schneider, W., & Maasen, S. (1998). *Mechanisms of visual attention: A cognitive neuroscience perspective.* Hove, UK: Psychology Press.

Schwartz, N., Ellsworth, L., Graham, L., & Knight, B. (1998). Accessing prior knowledge to remember text: A comparison of advanced organizers and maps. *Contemporary Educational Psychology, 23,* 65–89.

Stiles-Davis, J., Kritchevsky, M., & Bellugi, U. (1988). *Spatial cognition: Brain bases and development.* Hillsdale, NJ: Lawrence Erlbaum Associates.

Treisman, A., & Gelade, G. (1980). A feature integration theory of attention. *Cognitive Psychology, 12,* 97–136.

Wolfe, J. (1994). Guided Search 2.0: A revised model of visual search. *Psychonomic Bulletin and Review, 1,* 202–238.

PART IV

TEACHER PREPARATION FOR JOURNEYS WITH GEOSPATIAL TECHNOLOGIES

CHAPTER 13

SPATIAL THINKING

The Key to Success in Using Geospatial Technologies in the Social Studies Classroom

Sarah Witham Bednarz
Texas A&M University

Robert S. Bednarz
Texas A&M University

Many geographers and other social science educators have noted the slow rate of adoption of geospatial technologies (GST) by geography/social studies teachers. Although the number of people using such technologies in modern society is growing quickly, they have not been adopted by educators at a rate commensurate with expectations. Since its development as a desktop application for professionals in a wide range of occupations in the early 1990s, geographic information systems (GIS), its supporting science, geographic information science (GIScience), and related spatial visualization tools such as remote sensing (RS), collectively referred to here as geospatial technologies or GST, have diffused slowly in educational contexts and then, largely only into select elite primary and secondary classrooms. In the United States, despite a decade of efforts by geographers and GIS/RS vendors, GST struggle to win a wider audience among those edu-

Digital Geography: Geospatial Technologies in the Social Studies Classroom, pages 249–270
Copyright © 2008 by Information Age Publishing
All rights of reproduction in any form reserved.

cators, characterized as *respectable early adopters* (Longley, Goodchild, Maguire, & Rhind, 2001), who serve as role models and opinion leaders for the majority of teachers.

Among social studies professionals, geography educators have justified the introduction of GST using three competing and yet complementary rationales: a) the educative rationale: geospatial technologies support the teaching and learning of geography; b) the place-based rationale: GST are ideal tools for the study of geographical problems at a range of scales; and c) the workplace rationale: geospatial technologies are essential tools for knowledge workers in the 21st century.

Unfortunately, these arguments have not appealed to large numbers of teachers. In his seminal study on the status of GIS in the United States, Kerski (2003) conducted a national survey of educators in the 1,520 high schools that had purchased GIS software by 2000. These high schools represent less than 8% of the estimated 19,998 public and private high schools in the United States (CCSSO, 2001). Kerski found that only half of the educators whose schools owned the software were using it and of that number, a mere 20% used it in more than one lesson in more than one class. He also found that teachers thought the software was complex, lacked time to develop curriculum that used GIS, and had little technical support for training and software maintenance.

The slow diffusion of GST remains perplexing and problematic in light of the opportunities these innovative tools offer to social studies education. A number of nontrivial barriers to implementation have been identified (Bednarz & Ludwig, 1997). Key constraints include unequal access to hardware and software, the unavailability of pre-processed data, and a lack of technical support. Education-related barriers include excessive demands on teachers' time, insufficient training in the technology (Bednarz & Audet, 1999), weak institutional support, and pedagogic issues related to curriculum, instruction, and assessment (Alibrandi, 2001; Bednarz, 2001; Kerski, 2003). An assessment of the first summer GIS training conducted by ESRI for teachers in 1998 revealed that even pioneer teachers who understood GIS and had a sophisticated understanding of geography had a difficult time finding ways to incorporate GIS into their curriculum (Bednarz, 2004).

GST's failure to be widely adopted in classrooms in the United States may also be attributed to inadequate research concerning its effectiveness in promoting significant learning in social studies and science (Baker & Bednarz, 2003). Another factor impeding widespread adoption may be the strategies used to incorporate GST into pre-collegiate education. Finally, because of the decentralized nature of the US education system, no concerted national plan has been developed to coordinate efforts. Thus, the action that has taken place can be characterized as small-scale, tentative, and relatively localized.

Many of these impediments are not unique to GST. Research on the implementation of technology in general reveals that there is little fundamental understanding of the role of technology in student learning. It is clear that generic technological literacy is necessary but not sufficient to implement technology-based teaching and learning in schools. It is also clear that little guidance exists to assist educators in using technology to teach specific content (Wallace, 2004). In the case of GST, social science educators are only beginning to understand the knowledge and skills teachers need in order to engage students with social studies through GIS, remote sensing, and other key visualization tools. And although some believe that the environment in which GST are implemented can support or undermine teachers' instructional effectiveness, existing research offers few, if any, recommendations to guide educational practice. Lastly, perhaps the most important barrier to the diffusion of GST in social studies education arises from educators' lack of understanding of spatial thinking, the fundamental concepts and cognitive processes that support, and are supported by, these powerful technologies.

This chapter focuses on spatial thinking and ways GST may—or may not—assist and encourage spatial thinking. We begin by setting the context and making the case for spatial thinking. The chapter continues with a discussion of a model to implement spatial thinking and GST in social studies classrooms based on the work of Shulman (1986) and Mishra and Koehler (2006). We conclude with preliminary observations from an ongoing NSF-funded program, *Advancing Geospatial Skills in Science and Social Studies*. Special attention will be given to observed relationships between teachers' understanding and use of specific strategies to enhance students' spatial thinking abilities and the use of GST, particularly GIS, Internet GIS (IGIS), and remote sensing.

WHAT IS SPATIAL THINKING?

Spatial thinking is defined as the knowledge, skills, and habits of mind to use concepts of space (such as distance, orientation, distribution, and association), tools of representation (such as maps, graphs, and diagrams), and processes of reasoning (such as cognitive strategies to facilitate problem-solving and decision-making) to structure problems, find answers, and express solutions to these problems (NRC, 2006). It is a collection of cognitive skills that allows individuals to use space to model the world, real and imagined, in powerful ways. Spatial thinking permits students to comprehend relationships and structures in multiple ways and to remember them, through a range of representations, in both static and dynamic forms. To paraphrase the National Geography Standards, spatial thinking can be framed through two questions: what do students know (about space and tools of representation) and what can they do (processes of reasoning) with what they know?

Spatial thinking has traditionally been a subject of interest among cognitive scientists, psychologists, and behavioral geographers. Researchers have found that spatial thinking expertise develops in specific disciplines and that its transfer from one area to another is not automatic. Research has also confirmed that spatial thinking develops uniquely in individuals. Most importantly, however, are the findings that spatial thinking can, and should, be learned by all (NRC, 2006).

In recent years, interest in spatial thinking has exploded with the advent of sophisticated visualization tools, GST, and increasing concern with the need to prepare a spatially literate workforce. The effect of this increased interest has been a National Research Council study on spatial thinking, an NSF-funded Spatial Intelligence and Learning Center focused on understanding spatial learning (http://www.spatiallearning.org), and calls for spatial literacy both in the United States (Goodchild, 2006) and in the United Kingdom through the government-funded Spatial Literacy in Teaching Centre of Excellence (http://www.spatial-literacy.org).

The Case for Spatial Thinking

Spatial thinking is important in a variety of contexts and is used in every day life as well as in academic and workplace settings. The authors of the NRC report *Learning to Think Spatially* (NRC 2006) differentiated among three types of spatial thinking: thinking in space, thinking about space, and thinking with space. Walking to school, taking a shortcut to avoid a traffic jam, playing a team sport such as soccer, or packing a suitcase—actions that are performed in space—all require spatial thinking in a real-world context. This context can be referred to as the "geography of life spaces" (NRC, 2006). Success in a wide variety of vocational and academic pursuits such as air traffic control, medicine, urban planning, engineering, and the geosciences depend on this skill. The second type of spatial thinking, thinking about space, is typically employed to help individuals understand how the world works and its context is termed "the geography of physical spaces." A student who understands about physical space would comprehend the diffusion process of a new idea, culture trait, or even a disease. The third type of spatial thinking, thinking with, space is more abstract but perhaps the most powerful form of spatial thinking. Spatializing non-spatial data or using space as an organizing framework to conceptualize problems and make decisions is a very effective cognitive strategy. This third context, "the geography of intellectual spaces," is the least understood yet perhaps most generative context for spatial thinking. Although the authors of *Learning to Think Spatially* (NRC, 2006) distinguished among these three contexts, they recognized that these forms of spatial thinking are interrelated: thinking in space promotes thinking about space, while the use of space as a cognitive strategy adds power and aids in thinking in and about space.

Success in spatial thinking is correlated with success in school. Spatial thinking plays a major role in generic learning, remembering, and problem-solving (Golledge & Stimson, 1996). Spatial thinking can facilitate encoding and recalling information and communicating both spatial and non-spatial relationships (Kitchin & Blades, 2002; Kulhavey & Stock, 1996; NRC, 2006; Verdi, 2002). Research suggests, for example, that readers remember information both verbally and visually. This "dual encoding" or conjoint retention is enhanced when readers are encouraged to link what they read to images such as maps, concept maps, and other graphic representations (Paivio, 1986). The process of producing spatial representations such as graphs, maps, and flowcharts create schemas that link related items of information and provide a means to search one's memory. Linking *what* with *where*, and reading and thinking *with* and *through* visual representations make content more understandable and memorable (Liben, 2001).

Of course, spatial thinking is crucial to the discipline of geography. The National Geography Standards identify the essence of spatial thinking as "the spatial perspective," one of geography's two key perspectives, defining it as "The essential issue of 'whereness'—embodied in specific questions such as, Where is it? Why is it there?" (National Geography Education Standards Project, 1994, p 57). Spatial thinking, however, is important in other academic social sciences and the humanities where "the spatial turn" is evident, particularly in sociology, economics, and history (Cosgrove, 2004).

In K–12 social studies, spatial thinking supports a significant amount and variety of learning. Maps, graphs, images, diagrams, models, and visualizations describe and explain the functions, structures, relationships, and operations of various phenomena across the social sciences (Bednarz, Acheson, & Bednarz, 2006). Often, inherently non-spatial data are spatialized and expressed graphically to aid in analysis and comprehension. For example, it is difficult to detect trends and patterns in demographic data. However, when the data are put into a graphic representation such as a population pyramid, the gender and age structure of the population become clearer and inferences about the place and its society are possible. For example, historians use timelines to help students understand the sequence of, and intervals between, important events.

Spatial thinking also supports one of the key purposes of social studies education: "... to help young people develop the ability to make informed and reasoned decisions for the public good as citizens of a culturally diverse, democratic society in an interdependent world" (NCSS, 1994, p. 3). One can argue that the ability to apply spatial thinking to understand maps, images, and GST intelligently and critically is becoming a requirement to participate effectively as a citizen in modern society (Bednarz & Acheson, 2003). Because of the widespread availability of GST, especially through Internet sources, geographic representations have become ever more common in the daily lives of a larger number of people. Being able to access information and visualize conditions in real time can make students and others cognizant of issues and conditions, the precursor to action and deeper study. Today there is a plethora of web-based sources that provide real or near real-time data that can be used to engage spatially literate students in civic and political activities. Research indicates that when young people have opportunities to discuss issues, they tend to demonstrate greater interest in politics, improved critical thinking and communications skills, more civic knowledge, and a greater propensity to discuss such issues out of school (Center for Information and Research on Civic Learning and Engagement, 2003). Knowledge is power, indeed.

Finally, spatial thinking serves as the foundation for using and understanding GST. The U.S. Department of Labor has identified geospatial technologies as one of three key twenty-first century growth industries and has allocated more than six million dollars since 2004 to meet critical workforce shortages in the field (U.S. Department of Labor, 2007). Workplace applications of GST are outpacing our ability to produce qualified technicians, managers, and users. Although much of this training is highly technical, recognition of the fundamental role of spatial thinking in the context of GST is growing. In order to use GST to think effectively, students must: a) know what spatial thinking is and understand its key concepts (propositional knowledge—displayed in Figure 13.1); b) know how to apply spatial thinking (procedural knowledge—the skills and cognitive strategies associated with how to think spatially); and c) know when to use this powerful strategy (conditional knowledge—the habit of mind and capabilities to use it). The growth of GST and its increasing use places a new responsibility on social studies and other educators to prepare students for workplace and life roles which require them to think spatially.

Despite its importance, spatial thinking is not an explicit part of social studies curricula in elementary, middle, high school or college classrooms in the United States. Most students are not competent spatial thinkers. An analysis of the 2001 National Assessment of Educational Progress (NAEP) geography exam reveals that at every level (grades 4, 8, and 12) test items

Spatial Relations	Processes Used in GST
Recognizing spatial distribution and spatial patterns	Constructing gradients and surfaces
Identifying shapes	Layering
Recalling and representing layouts	Regionalizing
Connecting locations	Decomposing
Associating and correlating spatially distributed phenomena	Aggregating
Comprehending and using spatial hierarchies	Correlating
Regionalizing	Evaluating regularity or randomness
Comprehending distance decay and nearest neighbor effects in distributions (buffering)	Associating
Wayfinding in real world frames of reference	Assessing similarity
Imagining maps from verbal descriptions	Forming hierarchies
Sketch mapping	Assessing proximity (requires knowing location)
Comparing maps	Measuring distance
Overlaying and dissolving maps (windowing)	Measuring directions
	Defining shapes
	Defining patterns
	Determining cluster
	Determining dispersion

Figure 13.1. Key concepts and spatial thinking skills.

that require students to use and interpret maps (that is, to think spatially) were the most challenging. Additional research indicates that students find spatial thinking the most challenging component of GIS use; even when they can produce maps, they cannot interpret them effectively or use them to solve problems or make generalizations (Baker, 2002). Few social studies teachers are prepared, through their pre-service education, professional development, or published materials, to teach students the cognitive strategies associated with spatial thinking. In fact, only now are educators recognizing that spatial thinking is an important component of student learning.

ADVANCING GEOSPATIAL SKILLS IN SCIENCE AND SOCIAL SCIENCE

As noted earlier, GST have been slow to diffuse into social studies classrooms. After nearly 15 years of working with science and social studies teachers to adopt GST, we have become convinced that a fundamental barrier to GST's diffusion is teachers' lack of understanding of the nature and importance of spatial thinking; that is, the cognitive skills underlying these powerful technologies. In order to explore this issue, we proposed and received funding from the National Science Foundation for the program *Advancing Geospatial Skills in Science and Social Science* (AGSSS). Until now, there has been no application of spatial thinking research in psychology, cognitive science, and geography to develop social studies curriculum or improve instruction in spatial analysis and problem solving. It is a new, emerging field of interest with broad implications for a number of domains. AGSSS is a first step in the exploration of this area.

AGSSS connects geospatially-skilled graduate students, termed Fellows, with science and social studies teachers, grades 6–12, in a collaborative two-year cycle to enhance teachers' and students' spatial-thinking knowledge and skills. The remainder of this chapter focuses on our experiences while working with four high school social studies teachers and their students. All four teachers, self-identified geography specialists, work closely, as a team, to coordinate curriculum and activities for more than 500 grade 10 students enrolled in World Geography in a low-density urban setting in south central Texas.

The AGSSS program has five overall goals: a) to develop and test a set of training experiences for geospatially-skilled graduate students to prepare them to enhance their own spatial thinking and that of teachers and students; b) to prepare graduate students with the knowledge and skills to work collaboratively with teachers to enhance spatial thinking as supported by GST; c) to develop and test a set of training experiences for teachers to prepare them to use GST; d) to develop and test a set of training experiences to assist teachers in conducting research in their classrooms about student spatial thinking; and e) to share teacher-research findings within the AGSSS partnership and other communities thereby building a foundation of practical, classroom-based findings regarding spatial thinking. As the program has evolved, the goals related to spatial thinking have focused on three questions: a) What is the nature of spatial thinking in classroom settings? b) What practical, classroom-based strategies can be used to develop spatial thinking? and c) What is the role of spatial thinking in the implementation of GST?

The first year of the project (2005–2006) began with a one-day introduction for participating teachers and graduate students followed by one week of intensive training for the graduate students that prepared them to work in classrooms, to understand the social studies curriculum better, and to make them more aware of their own spatial thinking skills and abilities. This training was followed by bi-weekly seminars throughout the school year. As the year progressed, the teachers, principal investigators (all GST-skilled geography professors), and Fellows worked to spatialize the existing curriculum by enhancing the spatial nature of traditional class activities and by introducing technology-supported student activities to complement existing teacher-generated lesson plans. New, spatially rich curriculum elements were introduced as well, including activities from *Mapping Our World* (Malone, Palmer, & Voigt, 2004); ISS EarthKAM Missions, and geography-focused Google Earth activities. In June 2006, all teachers and Fellows participated in a week-long, co-learning, GST and spatial thinking training session that was catalytic and significant for all involved. The summative experience allowed participants to solidify their knowledge, to reflect on their practices, and to make generalizations about what worked—and what did not work—in spatial thinking and GST. These "lessons learned" are described in detail and form the core of the "Preliminary Conclusions" section of this chapter.

During this academic year (2006–2007), AGSSS Fellows and teachers are introducing more explicit and spatially-rich curriculum and practices. The process is one of revising and refining the initial units, expanding existing activities to make them more spatial, and implementing new activities. The focus on spatial thinking has been refocused and action research projects have been initiated in order to help us understand the nature of spatial thinking taking place in the classroom.

In order to comprehend and theorize the complex interactions occurring among geographers, graduate students, teachers, and their students, we have turned to the work of Shulman (1986, 1987) and Mishra & Koehler (2006). Their work provides a very useful conceptualization of how teachers typically adopt technologies and introduce them into their teaching.

DEVELOPING A CONCEPTUAL FRAMEWORK: CONTENT, PEDAGOGY AND TECHNOLOGY

The work of Shulman (1986, 1987) has been enormously influential in explaining how teachers understand and balance the competing dimensions of content and pedagogy in their teaching. Teaching requires *content knowledge*: knowing *what* to teach. Content knowledge is knowledge about subject matter and how it is practiced: in the context of the social studies it is understanding the key concepts and ideas, structure, and organization of social science disciplines such as geography. Teaching also requires knowledge about how to teach: *pedagogical knowledge,* knowledge about curriculum, lesson planning, assessment, and instructional strategies. Content knowledge and pedagogical knowledge may be competing goals, and thus, teacher-preparation courses may be more focused either on the acquisition of content knowledge or on the development of pedagogical skills. Even today, debates on teacher preparation reflect this either-or model; one camp believing that teachers need only a deep content knowledge to be successful in the classroom while the other holds that teachers require only a repertoire of sound instructional strategies to be effective (Bednarz, Stoltman, & Lee, 2004).

Shulman recast this recurrent debate by introducing the idea of *pedagogical content knowledge,* "the blending of content and pedagogy into an understanding of how particular topics, problems, or issues are organized, represented, and adapted to the diverse interests and abilities of learners, and presented for instruction" (Shulman, 1987, p. 8). Pedagogical content knowledge is knowledge about the best way to teach subject matter. Shulman described it as, "an understanding of what makes the learning of specific topics easy or difficult: the conceptions and preconceptions that students of different ages and backgrounds bring with them." It is "the most useful forms of representation of those ideas, the most powerful analogies, illustrations, examples, explanations and demonstrations—in a word, the ways of representing and formulating a subject that makes it comprehensible to others" (Shulman, 1986, p. 9). Pedagogical content knowledge develops as a teacher moves from novice to expert through a process of reasoning and experience-building from planning, teaching, adapting instruction, and reflecting on classroom experiences, that is, through a carefully coordinated series of learning experiences. The concept has been incorporated into almost all scholarship on teaching, learning, and teacher education since it was introduced and has helped unite what was previously seen as two separate domains. Mishra and Koehler (2006) used a graphic to show pedagogical content knowledge joining and uniting content and pedagogical knowledge (Figure 13.2).

Certainly the idea of content knowledge, pedagogical knowledge, and pedagogical content knowledge has been useful in articulating reasons for the slow diffusion of GIS and other GST into social studies classrooms (Bednarz & van der Schee, 2006). Many social studies teachers do not have an understanding of the spatial perspective that is key to geography and spatial thinking. This lack of content knowledge means that many teachers have limited pedagogical content knowledge, defined in this instance as knowledge about the best way to teach subject matter. The result is that few teachers recognize the potential opportunities GST offer to teach social studies content and skills.

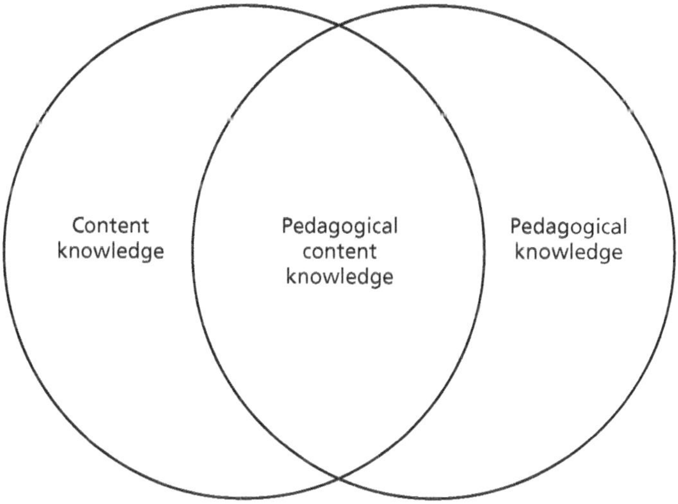

Figure 13.2. The two domains of pedagogical knowledge and content knowledge joined by pedagogical content knowledge (After Mishra and Koehler, 2006).

In addition, some social studies teachers lack the pedagogical knowledge required to teach effectively with technology. Hughes (2005) followed Shulman's conceptualization of teacher knowledge to examine the nature of teachers' learning during technology professional development activities and the extent to which their subsequent technology-supported pedagogy was innovative. She found that the power to develop innovative technology-supported pedagogy lay in the teacher's interpretation of the newly learned technology's value for supporting instruction and learning in the classroom and that learning experiences grounded in content-based, technology examples were most successful. Furthermore, she found that teachers with less professional knowledge (e.g., pre-service, novice or non-subject specialists) and/or less intrinsic interest in identifying uses for technology required guided or collaborative, content-specific technology learning opportunities, whereas teachers with more professional knowledge (e.g., veteran or subject specialists) were able to develop innovative technology-supported pedagogy by bringing their own learning goals to bear during professional development activities.

Although this research provided us with some ideas about potentially effective strategies for the AGSSS program, there was a missing component to the model. Mishra and Koehler (2006), who extended Shulman's work, added the third essential component needed by teacher's today, knowledge about technology. Technology encompasses both older devices such as overhead projectors and whiteboards and more recent digital technologies such as video, computers, and the Internet. Technology knowledge includes the

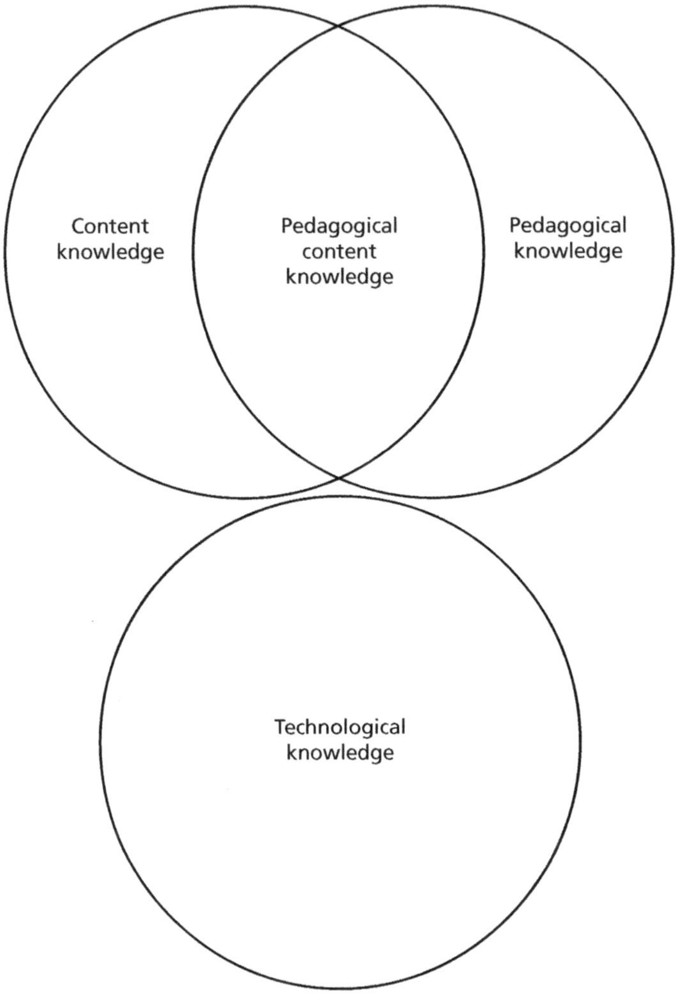

Figure 13.3. Technology as a separate domain of teacher knowledge (After Mishra and Koehler, 2006).

skills to operate hardware and to use and manage software (including GST). As Mishra and Koehler pointed out, technology knowledge has, until recently, been conceptualized as a separate domain, much as content and pedagogical knowledge was in the past (Figure 13.3). They are critical of most current research, theory development, and teacher training in educational technology because it conceptualizes technology as a body of knowledge and skills separate from content and pedagogical domains.

Instead, Mishra and Koehler propose a more complex, integrative model of teacher knowledge and practice (Figure 13.4) that unites con-

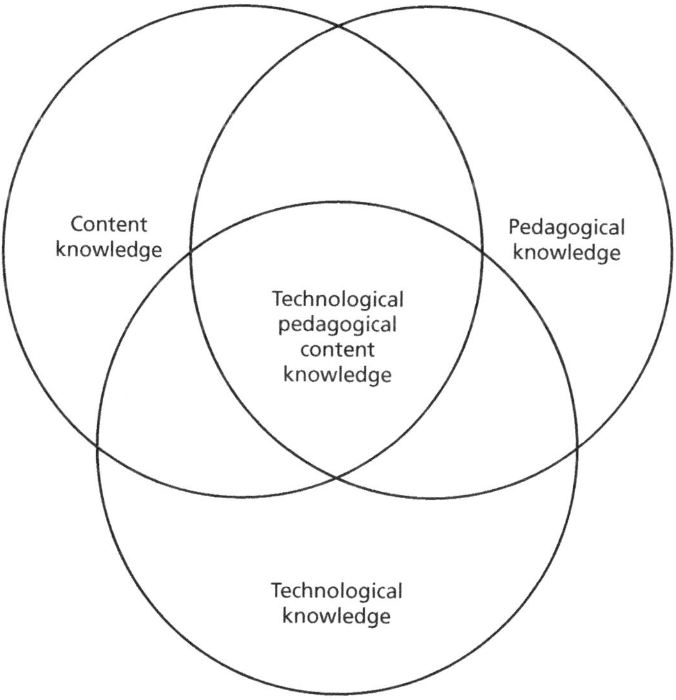

Figure 13.4. The intersection of technology, pedagogy, and content knowledge producing three new types of teacher knowledge: technological content knowledge, technological pedagogical knowledge, and technological pedagogical content knowledge (After Mishra and Koehler, 2006).

tent, pedagogy, and technology into a comprehensive framework. In addition to content knowledge, pedagogical knowledge, and technology knowledge, their model identifies three additional knowledge types that develop at the intersections of technology, content, and pedagogy, namely technological content knowledge, technological pedagogical knowledge, and technological pedagogical content knowledge. Technological content knowledge, or "knowledge about the manner in which technology and content are reciprocally related" is an understanding of the ways in which content is changed, enhanced, or altered by the use of a specific technology (Mishra & Koehler, 2006, p. 1028). In the context of GST, a teacher who sees how learning geography is changed by using GIS to compare two maps' data layers has technological content knowledge. Technological pedagogical knowledge, knowing how to teach with technology, is "the ability to choose a tool based on its fitness, strategies for using the tool's affordances, and knowledge of pedagogical strategies and the ability to apply those strategies for use of technologies" (Mishra & Koehler, 2006, p. 1028).

In a social studies classroom, a teacher who knows how to manage his or her class in a computer lab, how to assign group work for some technology-related tasks, and how to engage students in new and unfamiliar classroom settings might be said to have technological pedagogical knowledge. The third innovative knowledge type proposed in this framework represents the overlay of all types of teacher expertise related to content, pedagogy, and technology. Mishra and Koehler (2006) explained:

> Technological pedagogical content knowledge is the basis of good teaching with technology and requires an understanding of the representation of concepts using technologies; pedagogical techniques that use technologies in constructive ways to teach content; knowledge of what makes concepts difficult or easy to learn and how technology can help redress some of the problems that students face; knowledge of students' prior knowledge and theories of epistemology; and knowledge of how technologies can be used to build on existing knowledge and to develop new epistemologies or strengthen old ones. (p. 1029)

Simply put, teachers who possess technological pedagogical content knowledge know what technology is best to teach specific content and how to implement that technology so all students learn. This type of teaching requires the development of a sophisticated and integrated understanding of the relations among the three components of the model.

Preliminary Conclusions from AGSSS

Mishra and Koehler's framework explaining the relationships among content, pedagogy, and technology provides a method for examining and assessing the preliminary results of the AGSSS program, particularly for understanding the processes through which teachers adopt and use new ideas related to technology. The model has also offered a better understanding of observed classroom events, changes in teachers' and Fellows' knowledge about spatial thinking, and intended and unintended consequences. We devote the final section of this chapter to a discussion of four components of the framework: content knowledge, pedagogical content knowledge, technological pedagogical knowledge, and technological content pedagogical knowledge.

Content Knowledge

Despite intensive and repeated efforts to explain spatial thinking to the social studies teachers in the AGSSS program, it was not until teachers completed the co-learning week-long training in GST and spatial thinking held at the end of the first year that they reported "finally beginning to get it." Teacher awareness of spatial thinking and understanding of its importance to the discipline of geography through "the spatial perspective" was (and remains) slow to develop. The teachers had not included concepts of space, methods to represent space through maps and graphs, or spatial thinking processes as part of their conceptualizations of World Geography's subject matter. Only slowly and incrementally, as Fellows worked with them to spatialize their curriculum, did the teachers begin to understand how spatial thinking could organize and enhance World Geography content.

This result is not surprising. Most educators must, and probably should have to, be persuaded of the value of incorporating an innovation such as spatial thinking into their teaching methods as well as supported in doing so. The process of persuasion is complex and multidimensional and influenced by individual teachers' personal characteristics such as prior content knowledge, beliefs, and interests related to the teaching/learning process (Hughes, 2005; Murphy & Alexander, 2004). Teachers are gatekeepers of educational reform and innovation. They must be carefully guided and taught, a time-consuming and gradual process. Previous research on this topic reports the essentially conservative nature of teachers and their resistance to change. As Cohen (1995) pointed out:

> Teachers are the problem that policy must solve in the sense that their modest knowledge and skills are one important reason why most instruction has been relatively didactic and unambitious. But teachers also are the agents on whom policy must rely to solve that problem, for unless they learn much more about the subjects they teach and devise new approaches to instruction, most students' learning will not change. (p. 13)

Although we think this view of teachers is probably too pessimistic, given our experiences with AGSSS teachers, it does highlight the challenge of changing disciplinary conceptions—content knowledge—to include spatial thinking.

Pedagogical Content Knowledge

Pedagogical content knowledge is knowing how best to teach specific content to a range of learners. It is knowledge that develops over time and with experience, and as previously discussed, derives from a combination of deep content knowledge, an understanding of students' conceptions (and misconceptions) about the subject matter, and mastery of a range of instructional strategies. In the AGSSS context, we found that pedagogical content knowledge (i.e., understanding of why and how spatial thinking can help students acquire geography knowledge and skills) has also been slow to develop. The connections between spatial thinking, geography content, and appropriate instructional methods are developing, but only very slowly.

Again, this is not surprising. Teachers work hardest on what they value the most. As Hughes (2005) demonstrated in her research, interest in developing innovative, technology-supported pedagogy depended on the teacher's interpretation of the innovation's value for supporting classroom instruction and learning. If some of the AGSSS teachers do not understand the value of spatial thinking (content knowledge), they are unlikely to invest time and energy in finding effective ways to use it in their teaching (pedagogical knowledge), let alone to determine when, where, and how GST could help students learn particular concepts and skills (pedagogical content knowledge). If we can explain how spatial thinking supports geography learning more clearly and powerfully, we are confident that teachers will decide to incorporate it more quickly, and the rate of progress will increase. This is not to say progress is not being made; it is, but within the limitations outlined here. We still need to explore explicitly ways to better explain and communicate the power of spatial thinking to learning geography and other social studies and linking that to effective instructional strategies.

Technological Pedagogical Knowledge

The development of strategies to teach with technology and with spatial thinking has progressed more quickly and efficiently than expected. With the support of the AGSSS Fellows, the participating teachers have successfully managed classroom strategies in the social studies computer lab and integrated technologies such as a whiteboard, computer projector, and the Internet into their teaching. Initially teachers were reluctant to allow students to work on their own on GST-based computer activities (e.g., *Mapping Our World*); they insisted that they demonstrate activities and procedures before students began the activity. With time, teachers began to innovate and to request changes to the *Mapping Our World* activities to accommodate some of their learners. They also began experimenting with different student work schemes, such as letting students work on their own or in pairs. Thus, we perceive success in effecting change in teachers' technological pedagogical knowledge which is occurring at a rapid pace.

The wide range of hands-on, spatial thinking, GST products suggested by the teachers and developed by the Fellows are a second example of technological pedagogical knowledge success. Midway through the first year, we discovered that language was a barrier preventing spatial thinking. The students' lack of spatial vocabulary made it difficult for them to discuss spatial relationships and patterns. Students needed guided activities to help them understand key concepts such as shape, scale, association, and pattern and to use them frequently, in order to develop fluency. Teachers suggested that we construct a spatial vocabulary "flip book" that students could use to describe remotely-sensed images and general spatial relationships (see Figure 13.5). These laminated, spiral-bound books sit at the side

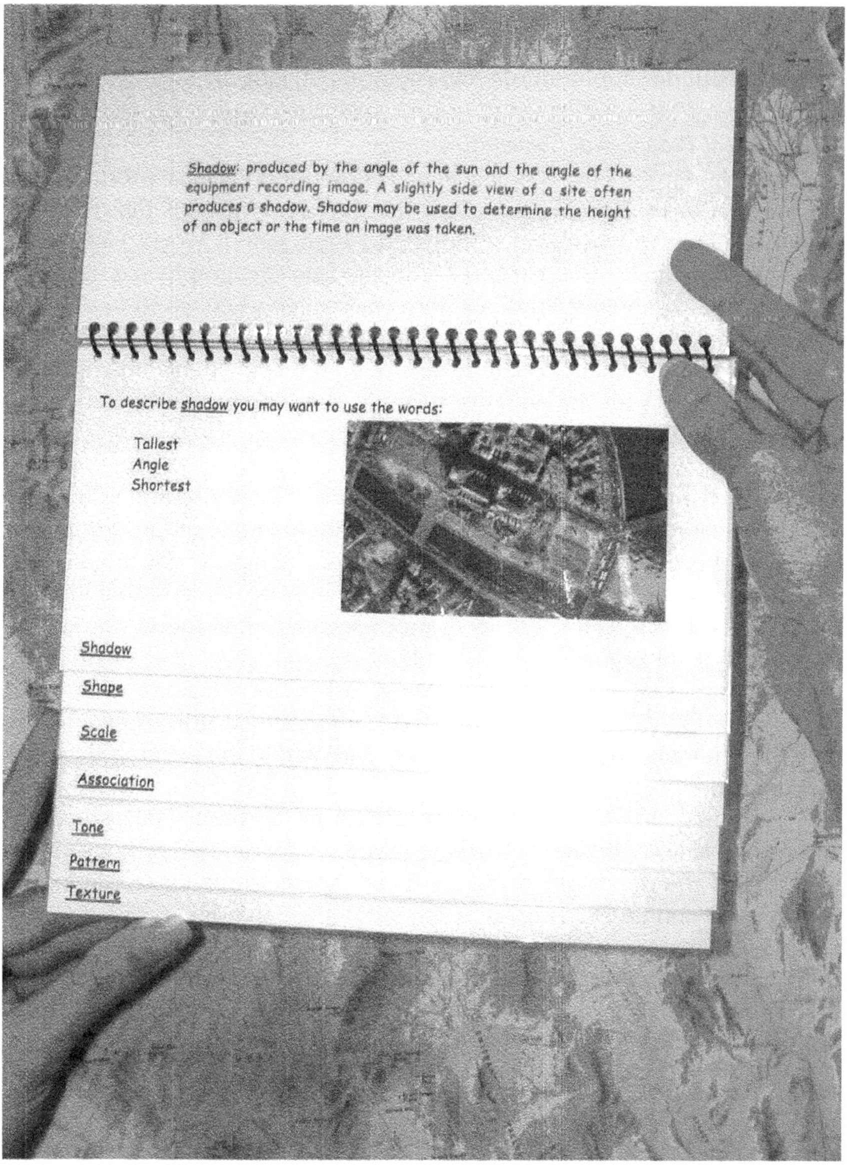

Figure 13.5. Evidence of the development of technological pedagogical knowledge: the spatial vocabulary flip book.

of the computer when students work in the computer lab and help students to develop their language of space. A second product teachers requested were software-specific "placemats or cheat sheets," guides to the key buttons, functions, and procedures of software. These also sit alongside computers providing students with the assistance they need to become independent users of the technology. These artifacts are evidence of teachers' successful development of technological-pedagogical knowledge.

Technological Pedagogical Content Knowledge

The collaborative efforts of teachers, Fellows, and PIs have produced initial observations that begin to answer the question, "What is the nature of spatial thinking in classroom settings?" In high school World Geography classes, student ability to use a GST—GIS—to learn geography was inhibited by a lack of a vocabulary. Students were challenged when asked to describe world regions, spatial patterns (e.g., clusters of volcanoes at the interface of tectonic plates), or spatial relationships (e.g., the coincidence of volcanoes and earthquakes with the edges of tectonic plates). This became apparent during a Google Earth activity late in the first year when students were asked to "zoom in to" a location "at the edge" of Cairo and the surrounding desert. We wanted students to interpret this very distinctive line as the border between urban land use and desert. Students understood "zoom in." However, the concept "edge" confused them. "What do you mean, 'the edge?'" they asked. This event and others like it made us aware that we were asking students to perform tasks (pattern seeking, description, navigation from place to place) that were completely new to them. Not surprisingly, students were unable to accomplish these tasks until they were introduced to the appropriate spatial concepts. They also needed guided practice to help them understand how these new ideas related to the geographic representations they were using, and how they should proceed to use the technology (zooming in; zooming out; maintaining a frame of reference) appropriately and effectively. These classroom experiences were an important part of the development of the team's spatial thinking technological pedagogical content knowledge. Only after this experience did the four teachers, Fellows, and PIs begin to understand that explicit vocabulary training, pedagogical tools (i.e., the placemats and flip books), and guided practice using GST were necessary to help the more than 500 World Geography students learn to think spatially.

CONCLUSIONS

This chapter began by documenting the slow diffusion of GST, particularly GIS, into U.S. classrooms. We suggested a wide range of reasons for this sluggish rate of adoption, including teachers' unfamiliarity with spatial thinking—the foundation of GST. We went on to describe spatial thinking, making the point that although it is a ubiquitous part of daily life, academic success, and a growing number of workplace skills, it is also almost never taught explicitly in K–12 education. We then described the AGSSS program and ways participants are exploring the connections between spatial thinking and GST in classroom contexts. By applying the frameworks of teachers' knowledge devised by Shulman (1986) and Mishra and Koehler (2006), we interpreted preliminary results of the AGSSS program.

Before we began the AGSSS project, we suspected that a lack of appreciation for, and understanding of, spatial thinking might be contributing to the slow implementation of GST. Our recent experiences and the preliminary results reported in this chapter only reinforce our belief in the close positive relationship between familiarity with spatial thinking and the development of the technological pedagogical content knowledge required to effectively infuse GST in social studies classrooms. To conclude we wish to make three observations and to discuss the implications of these observations for social studies educators. The first observation is that educational change does not happen quickly, especially if it is not mandated or well-supported. As we have seen, and other research confirms, it takes considerable time to develop content knowledge, pedagogical knowledge, and, eventually, pedagogical content knowledge. The process of reconceptualizing curriculum, building pedagogical skills, and learning to teach content effectively is challenging and occurs incrementally even in situations where teachers are given support. Our second observation is that some educational changes take place more quickly than others. For example, we observed that teachers developed technological pedagogical knowledge fairly readily. We can speculate that this type of change may be easier and quicker to develop because although it required teachers to learn something new, it did not force them to break an existing pattern of behavior or habit. A second reason contributing to the relative ease by which the four geography teachers gained technological pedagogical knowledge could be the benefits they perceive in managing their students efficiently and effectively using more appropriate instructional strategies. The prospect of unruly students is a strong incentive for teachers to be creative and to innovate. Our third observation is that spatial vocabulary and concept understanding are key to spatial thinking and thus to the development of technological pedagogical content knowledge. As students have become more fluent in the language of space, they have become more adept spatial thinkers. The same is true for the AGSSS partner teachers. As they have become aware of the connection between language and conceptual understanding, they have begun to be more explicit about their use of spatial language and terms. As a result, they report improved student performance on written assignments requiring students to use spatial concepts and in the level and richness of students' questions while they are engaged in GST-supported activities.

These observations have implications for social studies educators, curriculum developers, researchers, and others interested in GST-supported education. While we remain cautiously optimistic about the value of GST to support teaching spatial thinking and geography knowledge and skills, we suggest that this relationship be re-ordered and that spatial thinking be the necessary precursor to (or at least explicitly co-taught with) in the implementation of GST in social studies classrooms. We imagine that as social studies teachers develop their technological pedagogical content knowledge in spatial thinking and GST, they will prepare their students for GST by seeding them with the needed spatial thinking skills. A second implication is that there is a need for careful exploration of the kinds and levels of vocabulary support students need to be more successful spatial thinkers. This is especially true for the development of instructional materials using GST. Students must be provided with scaffolding if they are to perform the kinds of spatial thinking GST demand. We are also concerned with the implications for teacher preparation. If pre-service teachers are not given the opportunity to hone their own spatial thinking skills (let alone skills in using GST), it is highly unlikely that they will be able to develop the technological pedagogical content knowledge required to exploit the capabilities of GST for social studies education.

NOTE

Much of the material in this chapter was presented at the 2006 ESRI Education Users Conference; the conference *Changing Geographies: Innovative Curricula* sponsored by the Institute of Education, University of London (2007); the National Council for Geographic Education Annual Meetings (2006); and at the Seventh National Science Foundation Graduate Teaching Fellows in K–12 Education Project Meeting.

REFERENCES

Alibrandi, M. (2001). Making a place for technology in teacher education with geographic information systems (GIS). *Contemporary Issues in Technology & Teacher Education, 1*(4), 483–500.

Baker, T. R. (2002). The effects of Geographic Information System (GIS) technologies on students' attitudes, self-efficacy, and achievement in middle school science classrooms. (Doctoral dissertation, The University of Kansas, 2002).

Baker, T. R., & Bednarz, S.W. (2003). Lessons learned from reviewing research in GIS education. *Journal of Geography, 102*(6), 231–233.

Baker, T. R., & White, S. H. (2003). The effects of GIS on students' attitudes, self-efficacy, and achievement in middle school science classrooms. *Journal of Geography, 102*(6), 243–254.

Bednarz, S. W. (2001) Thinking spatially: Incorporating geographic information science in pre and post secondary education. In L. Houtsonen & M. Tammilehto (Eds.), *Innovative Practices in Geographical Education* (pp. 3–7). Helsinki: Proceedings of Helsinki Symposium, IGU Commission on Geographical Education.

Bednarz, S. W. (2004). Geographic information systems: A tool to support geography and environmental education? *GeoJournal 60*(2), 191–199.

Bednarz, S. W., & Acheson, G. (2003). Learning to be a citizen in post 9/11 United States: What role for geography? *Proceedings of the Commission on Geographic Education, Geography and Citizenship Education.* London: Institute of Education.

Bednarz, S.W., & Audet. R. (1999). The status of GIS technology in teacher preparation programs. *Journal of Geography, 98*(2), 60–67.

Bednarz, S. W., & Ludwig, G. (1997). Ten things higher education needs to know about GIS in primary and secondary education. *Transactions in GIS, 2*(2), 123–133.

Bednarz, S. W., Stoltman, J. P., & Lee. J. (2004). Preparing geography teachers in the United States. *International Research in Geographical and Environmental Education, 13*(2), 176–183.

Bednarz, S. W., & van der Schee, J. (2006). Europe and the United States: The implementation of geographic information systems in secondary education in two contexts. *Technology, Pedagogy, and Education, 15*(20), 191–205.

Bednarz, S. W., Acheson, G., & Bednarz, R.S. (2006). Maps and map learning in social studies. *Social Education, 70*(7), 398–404.

Center for Information and Research on Civic Learning and Engagement. (2003). *The civic mission of schools.* New York: Carnegie Corporation.

Cohen, D. K. (1995). What is the system in systemic reform? *Educational Researcher, 24*(9), 11–17.

Cosgrove, D. (2004, Fall). Landscape and landschaft. *German Historical Institute Bulletin, 35*, 57–71.

Council of Chief State School Officers (CCSSO). (2001). *Profiles of state education systems for use with NAEP.* Retrieved January 17, 2007, from http://www.ccsso.org/projects/Accountability_Systems/State_Profiles/

Golledge, R. G., & Stimson, R. J. (1996). *Spatial behavior.* New York: The Guilford Press.

Goodchild, M. (2006). The fourth R? Rethinking GIS education. *ArcNews, 28*(3), 5–7.

Hughes, J. (2005). The role of teacher knowledge and learning experiences in forming technology-integrated pedagogy. *Journal of Technology and Teacher Education 13*(2), 277–302.

Kerski, J. J. (2003). The implementation and effectiveness of geographic information systems technology and methods in secondary education. *Journal of Geography, 102*(4), 128–137.

Kitchin, R., & Blades, M. (2002). *The cognition of geographic space.* London: I.B. Taurus.

Kulhavey, R. W., & Stock,W. A. (1996). How cognitive maps are learned and remembered. *Annals of the Association of American Geographers, 86*(1), 123–145.

Liben, L. (2001). Thinking through maps. In M. Gattis (Ed.), *Spatial schemas and abstract thought* (pp. 45–78). Cambridge, MA: MIT Press.

Longley, P. A., Goodchild, M. F., Maguire, D. J., & Rhind, D. W. (2001). *Geographic information systems and science.* New York: Wiley.

Malone, L., Palmer, A. M., & Voigt, C. L. (2004). *Mapping our world: GIS lessons for educators.* Redlands, CA: ESRI Press.

Mishra, P., & Koehler, M. J. (2006). Technological pedagogical content knowledge: A framework for teacher knowledge. *Teachers College Record, 108*(6), 1017–1054.

Murphy, P. K., & Alexander, P. A. (2004). Persuasion as a dynamic multidimensional process: An investigation of individual and intra-individual differences. *American Educational Research Journal, 41*(2), 337–363.

National Council for the Social Studies. (1994). *Curriculum standards for the social studies: Expectations of excellence.* Washington DC: National Council for the Social Studies.

National Geography Education Standards Project. (1994). *National Geography Standards 1994: Geography for Life.* Washington DC: National Geographic Society Committee on Research and Exploration.

National Research Council (NRC) (2006). Committee on Support for Thinking Spatially. (2006). *Learning to think spatially.* Washington DC: National Academies Press.

Paivio, A. (1986). *Mental representations.* New York: Oxford University Press.

Shulman, L. (1986) Those who understand: Knowledge growth in teaching. *Educational Researcher 15*(2), 4–14.

Shulman, L. (1987) Knowledge and teaching: Foundations of the new reform. *Harvard Educational Review 57*(1), 1–22.

U.S. Department of Labor. (2007). Local solutions with national applications to address geospatial technology industry workforce needs. Retrieved January 12, 2007, from http://www.doleta.gov/BRG/indproof/geospatial.cfm.

Verdi, M. P. (2002). Learning effects of print and digital maps. *Journal of Research on Technology in Education, 35*(2), 290–303.

Wallace, R. M. (2004). A framework for understanding teaching with the Internet. *American Educational Research Journal, 41*(2), 447–488.

CHAPTER 14

EXAMINING THE TEACHER'S ROLE WHEN TEACHING WITH GEOGRAPHIC INFORMATION SYSTEMS (GIS)

Eui-kyung Shin
Northern Illinois University

Teachers are the primary instructional and curricular decision-makers in classrooms (Clandinin & Connelly, 1992; Thornton, 1991). When teachers incorporate new technologies into their curriculum and instruction, it is important how the technology is introduced and how efficiently teachers use it to help students achieve instructional goals (Budin, 1991; Olson & Clough, 2001). Since the successful implementation of technology in classrooms is heavily dependent on teachers' curricular-instructional decision-making, it is important to examine the teachers' role in using technology for classroom instruction. While there has been some research about the teachers' role in using technology in general (Blissett & Atkins, 1993; Budin, 1991; Koo & Harlin, 2001; Olson & Clough, 2001; Rasku-Puttonen, Etelapelto, Arvaja & Hakkinen, 2003; Tiene & Luft, 2001; Wang, 2001), the teacher's role in using Geographic Information Systems (GIS) technology specifically has not been extensively studied.

The use of GIS in K–12 classrooms has been promoted by many scholars in recent decades (Alibrandi, 2003; Baker & White, 2003; Bednarz, 2004; ESRI, 2000; Geography Education Standards Project, 1994; Kerski,

Digital Geography: Geospatial Technologies in the Social Studies Classroom, pages 271–290
Copyright © 2008 by Information Age Publishing
All rights of reproduction in any form reserved.

2001; Mason & Alibrandi, 1999; Shin, 2006; West, 2003), and some examples of GIS applications in classrooms have been provided (Alibrandi, 2003; Alibrandi & Sarnoff, 2006; Kerski, 2003; Shin, 2006). However, much of the research on using GIS has been focused on GIS intervention itself and students' learning rather than on the role of teachers when integrating GIS in classrooms. Therefore, the purpose of this study was to examine how a teacher adapted and integrated the GIS technology into a fourth grade social studies classroom, especially when GIS was introduced for the first time.

THE TEACHER'S ROLE IN USING TECHNOLOGY

When technology was first adopted in classrooms, many teachers feared to an extreme that one day computers would replace teachers (Budin, 1991), yet research has confirmed that the role of the teacher is critical to the success of instruction when technology is used. Unless the teacher combines technology with effective instructional processes, technology is unlikely to be used effectively (Rose & Winterfeldt, 1998). In emphasizing the importance of the teacher's role in instruction using technology, Olson and Clough (2001) stated, "Teachers translate curriculum into a form ready for classroom application and decide what, how, and why to learn.... The principles of effective teaching are not changed by the presence or absence of technology" (p. 11). They further explained that the success of teaching is based upon teachers' effective use of questioning and the ability to keep students mentally engaged.

Although how teachers practice in the classroom while using technology remains important, technology use itself seemed to change the teacher's role in instruction (Rasku-Puttonen et al., 2003). In a technology rich environment, the teachers' role becomes "coach," rather than "information dispenser" (Tiene & Luft, 2001; Wang, 2001). Teachers tend to spend less time on whole group instruction. The interaction between the teacher and students becomes more small-group oriented and focuses on "guiding" them rather than "teaching" them (Tiene & Luft, 2001). Therefore the use of technology can help to transform the classroom from "teacher-centered" and "curriculum-centered" instruction to more "student-centered" instruction (Milman & Heinecke, 2000; Wang, 2001).

In addition, the use of technology changes the dynamics of interaction between the teacher and students and among students. Despite concerns that the technology use would encourage less social interaction among students, research reported that students tended to interact more when they were working with computers because, in real classrooms, students were often asked to share a computer with their peers and encouraged to help each other in completing a task (Olson & Clough, 2001; Wang, 2001). These changes in the teachers' role and classroom dynamics have affected the technology training for teachers. Teachers not only need to be the experts in using technology, but also need to develop appropriate teaching styles (Doolittle & Hicks, 2003; Wang, 2001).

In the study of geography, the National Geography Standards acknowledge that GIS is a technological tool that can be used to enhance geographic skills more effectively (Geography Education Standards Project, 1994). However, it also noted that geography cannot be simply done by a computer system "because every machine requires an operator, someone who decides what is to be done and why, someone who can ask questions and answer" (Geography Education Standards Project, 1994, p. 256). In explaining the teachers' involvement when using GIS, Kerski (2003) explained the nature of GIS technology as "The tool itself has no answers-these come from human investigators" (p. 131). He emphasized the importance of teacher training for GIS integration.

Nonetheless, whether for technology in general or GIS specifically, how teachers use the technology tool is important. However, not much research has been conducted that focuses on how teachers use GIS in actual classrooms nor have many exemplary models of classroom teachers implementing GIS been shared.

THE TEACHER'S ROLE IN DECISION MAKING

As with any other instructional setting, the classroom teachers' understanding of the curriculum makes a significant difference in instruction and student learning (Shulman, 1986, 1987; Wiske, 1998). Historically, teachers have been seen as people who carry out the curriculum decisions made by policy makers and school district officers (Budin, 1991; Griffin, 1994). However, Thornton (1991) has theorized convincingly that teachers act as "curricular-instructional gatekeepers" by making professional decisions about which curriculum will actually be taught and selecting the best means for teaching it. Further, the lack of teachers' ownership is often reported as one of the primary reasons for unsuccessful restructuring of the curriculum (Nolan & Meister, 2000). Integrating GIS into the curriculum, therefore, requires teacher participation in both the design process and the implementation stage (Clandinin & Connelly, 1992).

The integration of technology is not an exception to the notion that teachers are "gatekeepers" who make instructional and curricular decisions. For example, Molebash (2002) explained that the philosophical beliefs of the instructor and the use of technology had an effect on one another. The instructors' beliefs in constructivism allowed them to integrate technology more effectively and vice versa. Doolittle and Hicks (2003) also provided the connection between pedagogy and technology integration. They argued that integrating technology in social studies education should follow pedagogical strategies to enhance students' learning, rather than using technology itself.

Although the teacher is one of the keys to successful infusion of GIS into a classroom setting, research that examines the teachers' role in using GIS is limited. Several studies on the use of GIS in K–12 classrooms have reported some level of success in student learning and the instruction provided (Alibrandi, 2003; Alibrandi & Sarnoff, 2006; Baker & White, 2003; Keiper, 1999; Shin, 2006; West, 2003; Wiegand, 2003; Wigglesworth, 2003). Additional research has reported the effectiveness of in-service and pre-service teachers' training (Gatrell, 2004; Kerski, 2003; Wilder, Brinkerhoff, & Higgins, 2003). Yet there has been no research specifically focused on how teachers teach with GIS or how they make their instructional and curricular decisions while using GIS technology. Therefore, the purpose of this study was to examine teacher decision-making and constructivist practice while using GIS technology.

METHODOLOGY

In this study, I used a qualitative design to explore the teacher's role in instruction and the teacher's decision making process in the use of GIS technology in a fourth grade classroom. The research centered on the interaction between the teacher and the students. A qualitative research design was chosen, since I hoped "to inform our deep understanding of educational institutions and processes through interpretation and narrative description" (Soltis, 1990, p. 249).

This research specifically uses the case study method. Case study is appropriate when the researcher examines a single entity of the phenomenon in-depth by collecting detailed information using a variety of data sources (Merriam, 1998). The use of multiple data sources ensures triangulation of data (Creswell, 1994; Merriam 1998), and the analysis of the data helps develop themes or patterns that propose implications to the field (Lincoln & Guba, 1985). Therefore, the rich description of the criteria for selecting the case, the setting, the GIS intervention, data sources and analysis methods are provided.

In this study, I took on the role of both participant and observer (Glensne, 1999; Spradley, 1980) by providing computer assistance for activities during classroom instruction, since the teacher was not familiar with the GIS technology. The teacher's lack of familiarity with GIS brought occasions when the students asked me some specific questions about GIS functions. On these occasions, I interacted with the students and answered their questions. However, the interaction between the students and me was minimal, and the teacher was in charge of instruction.

Criteria

This case study site was chosen based on several criteria. First, the school should represent a diverse group of students from a lower to middle class neighborhood. Second, the classroom should have an infrastructure of technology; however, not necessarily a fully equipped computer lab. Third, the students should reflect diversity in their academic ability. Fourth, the classroom teacher should be interested in learning and using new technology and possess some level of proficiency in using technology. However, she/he did not necessarily have prior experience using GIS. Lastly, the classroom teacher should be skilled in social studies instruction and classroom management.

In summary, the case study site needed to reflect a "typical" educational setting. The classroom needed to be in a lower to middle class neighborhood. The level of technology use at the school needed to be typical of many schools, with students having some experience using technology but with no experience using GIS. These criteria were applied to capture the best possible realistic view of the interaction between the teacher and students. The researcher believed that diverse participants working in a "typical" setting could provide better insights into practical application rather than an affluent community with a homogeneous student population. The researcher also wanted to work with a skilled teacher who was confident in managing the classroom. Based on these criteria, the fourth grade classroom of Ms. McNeill in Chantilly Elementary School[1] was chosen for the case study site.

School

This research took place in a fourth grade classroom at Chantilly Elementary School in a suburban neighborhood of Columbia, South Carolina. The socio-economic status of the students attending this school was low to middle class. According to the report card from the South Carolina Department of Education (2003), the total enrollment of this school was 592, and 58.7% of the students came from below the poverty level. The majority of the students scored below basic or scored at basic levels in both English and mathematics in the Palmetto Achievement Challenges Test (PACT).[2] Specifically, 20.8% scored below basic, 50.3% scored at basic,

26.5% scored at proficient, and 1.9% scored at the advanced level in English. In mathematics, 37.3% of the students scored below basic, 38.3% scored at basic, 14.6% scored proficient, and 9.7% of the students were at the advanced level.

Classroom and Equipment

The classroom used by these students was a portable classroom located in the school's back yard along with the other fourth and fifth grade portable classrooms. Students' desks were arranged in a horseshoe shape to allow for maximum interaction and eye contact between students. The classroom had an extensive display of student work on the walls.

There were originally two computers located near the front of the classroom. One of the computers was occasionally connected to the TV in front of the classroom. However, the second computer was old and slow and lacked enough memory space to meet the requirements for GIS software installation. Thus, the placement of the computers in the classroom was not appropriate for data gathering. For example, the two computers were placed on the same desk and too close to the students' desks. Therefore, the researcher brought two computers, one laptop and one desktop computer to the classroom, and those two computers were deliberately placed in the back of the classroom to allow students space to work. The computer that was located at the front was connected to the TV and was used as the demonstration station for the GIS activities.

The ArcView 3.2 software was used for the lessons in this project. This particular software was chosen for several reasons. First, it was donated by the software company, ESRI. Second, it allows for the display of multimedia components such as sound, graphics, and video clips. Third, the common GIS data format (mostly "shape" files) could be used with this software. Lastly, it is relatively easy to use among GIS software.

Teacher

The teacher, Ms. McNeill, was in her fourth year of teaching fourth graders. She majored in mathematics and had her M.A.T. degree in elementary education from a local university with her teaching certificate. She felt confident in her teaching and classroom management. In fact, she was elected as the teacher of the year in her third year of teaching. She was a member of the South Carolina Social Studies Council, but she did not have any membership or affiliation devoted specifically to geography education.

Although she did not regard herself as a computer expert, she had incorporated technology into her teaching. She had been using educational software such as Inspiration and the Internet to aid her teaching; however, she had no previous experience using GIS. She understood the importance of geography in social studies curricula and expressed her interest to implement GIS in her classroom. Although she showed a great deal of interest in GIS, she did not feel confident using GIS for instruction. Instead, she asked the researcher to help her learn GIS throughout the process and to provide technical help with GIS during her instruction.

Students

A total of eighteen students in this fourth grade class participated in this study: fifteen African-Americans and three Caucasians. Of the 18 students, there were fourteen females and four males. Ten students were on the free or reduced lunch program. According to their second grade PACT scores, one student scored below basic, five students scored at basic, five students scored at proficient, and one student scored at the advanced level in English.[3] In mathematics,[4] six students scored below basic, three students scored at basic, two students scored at proficient, and one student scored at the advanced level. Two students from this class participated in the speech program, and one student participated in the resource class. These students were scheduled to go to speech class or resource class during the social studies class, so the teacher made special arrangements for their group projects. For example, since April and Larry went to speech class, the teacher made arrangements for their group to use the computer either before or after the speech class. Since the teacher had incorporated computers and other technologies in her teaching, the students had previously worked with computers, such as educational software and the Internet.

During the instruction period, the students were divided into six different groups with three students in each group. These groups worked together in teams during small group activities. Ms. McNeill chose to group them at random by drawing their names from a box. Membership in these groups remained the same throughout the study. The students formulated a semi-agreeable system of working on the computer in groups in which they assigned group members to the roles of "clicker" (the person who controls the mouse), "manager" (or time keeper) and "writer" (the person who takes notes). This existing arrangement became useful, even though there were still some arguments that took place later.

The GIS Module

In order to introduce GIS to the students, a unit that is composed of four lessons was developed and implemented. The module is designed to move the students from learning the basic functions of GIS to more difficult and challenging tasks. When Ms. McNeill agreed to adopt this module, she was encouraged to make any necessary changes so it would fit into her social studies curriculum. To help the teacher, in case she needed assistance with the program, a simple one-page "cheat sheet" that explained buttons and functions with graphics was provided with the lesson plan.

In using the GIS module, no specific directions were given to the teacher. I met the teacher regularly to discuss the instruction; however, every attempt was made to allow the teacher maximum freedom in using this GIS module. During the meeting, I clarified issues and concerns that the teacher had about the GIS module and encouraged her to reconstruct the lesson if necessary. The classroom instruction was completed in thirteen class sessions and each session lasted between 40 and 75 minutes.

An overview of the GIS project is as follows. In the first lesson, a GIS activity dealing with growth patterns and changes in people and transportation in the City of Columbia was developed and implemented. In the second lesson, a GIS activity dealing with major modes of transportation and patterns of population distribution in the state of South Carolina over a period of time (1790, 1860 and 2000) was developed. In the third lesson, a GIS activity dealing with population growth and movement patterns in the United States was developed and implemented. Students were asked to compare the location of large cities in the United States in 1790, 1860, and 2000 and to connect to the development of transportation that people used during these time periods. In the fourth lesson, the GIS activity was an open-ended inquiry project. A list of data was given to the students and they were asked to decide on a topic of their choice based on their interest. Students were also asked to generate the maps based on the topic they chose and to make a class presentation of their work.

Data Gathering

In this study, data were collected using various sources to enhance its validity and to triangulate (Frenkel & Wallen, 1996). The various means of data gathering included the researcher's field notes, video and audio recording during instructions, interviews with the teacher, and students' work produced during class. Two video recorders were placed at the back of the classroom at different angles. These video recorders were used to record whole group instruction and GIS activities that were taking place at the two computer stations in the back of the classroom. These videotapes were transcribed and the data were used for discourse analysis. Audio recorders were used as well. The students were given one audio recorder per group during the group project to capture the discussion with the teacher and among themselves. These audiotapes were transcribed, and the data were used for discourse analysis as well.

The interviews with the teacher were conducted after each session to discuss the lesson and to prepare for the subsequent lesson. These interviews were recorded and transcribed. All the students' work, such as worksheets and maps produced using GIS during the instruction period, were gathered and used for data analysis. Finally, field notes were taken by the researcher during the instructional period and when interviewing the teacher after each session. After taking initial notes, more detailed notes were taken while watching the videotapes that recorded the sessions.

Data Analysis

The primary method of data analysis was discourse analysis. Transcripts of the interaction between the teacher and the students, interviews with the teacher and the researcher's field notes were analyzed to identify emerging themes. The researcher used a constant-comparative method (Glaser & Strauss, 1967) in coding and categorizing. First, the researcher read and re-read the data several times. The data were then coded based on what sort of role the teacher took during the instruction and the decisions that the teacher made regarding instruction. The coded data were classified into several categories and were examined to discern noticeable patterns that were developed into several themes presented below.

FINDINGS

The findings from this study confirm previous research that the role of the teacher is critical when technology is used for classroom instruction (Blissett & Atkins, 1993; Budin, 1991; Koo & Harlin, 2001; Milson, 2002; Olson & Clough, 2001; Rasku-Puttonen et al., 2003; Tiene & Luft, 2001; Wang, 2001). In this study, the teacher served as a model while demonstrating the use of GIS to the students and as a facilitator who prompted the students to connect to their prior knowledge and guided their thinking paths during the whole group and small group instruction. In addition, the teacher planned and directed what and how to introduce the content of the instruction while implementing the GIS technology, becoming the constructor for the curriculum used in the classroom.

Teacher as a Model for GIS Technology Use

When Ms. McNeill introduced each lesson, she first demonstrated the processes and the use of GIS functions in front of the students while asking some guiding questions. Since there was no written guideline about the GIS functions provided specifically to the students, the teacher's modeling was the only opportunity for the students to learn how to use the GIS technology. Although the teacher's modeling was the only source for learning GIS functions, the students seemed comfortable working with computers during the small group activities. They asked me a few questions, when they were not sure which button to click; however, they did not seem to be frustrated at all. For example, the final lesson required students to produce a map of the state or the United States. This was a challenging GIS skill that Ms. McNeill demonstrated while she verbally explained "what to click."

Ms. McNeill: This is what you do. This is what you're going to have to do to create your map. Let's say you want...Those are the households that have the most 15–17 year olds. What you do is you double click on "state." It comes up. It says "legend type." This is what you choose how you want it to be. And you go with "graduated color." Now it says "classification field." What you want to classify the state as... and that's where you look. And you slide down to "15–17." I want to know how many kids between 15 and 17. Which states have the most 15–17 year olds? And I don't like red. I want blue.

Students: Yeah

Ms. McNeill: And then I go to "apply." And then I go ahead and close this. And this is what the map looks like.

Students: Yeah good!

Ms. McNeill: And then I will make it large so you can see it. That's what it looks like. Now you will take this map and you will print it.

During the inquiry project, two groups (out of six) needed extensive help and asked me what to do; however, other groups were able to proceed by themselves with minor help from the researcher. It seemed that at least one of each group remembered what to do from the teacher's presentation and guided the rest.

Teacher as a Facilitator for the Geography Learning

Throughout the instruction, Ms. McNeill spent significant time encouraging the students to connect their prior knowledge and guiding their thinking paths. Even during whole group instruction, Ms. McNeill devoted a great deal of time and effort to connecting with their prior knowledge by bringing additional information and asking leading questions. For example, during the first session, when Ms. McNeill led a discussion comparing the size of Columbia in 1893 and 2000, she referred to the interstates that pass by Columbia, which the students used on regular basis, to guide them to connect to their daily lives. Ms. McNeill also mentioned the history of Columbia in terms of what people did for a living and how they traveled in order to help the students make connections between their prior knowledge and what they were going to learn.

Ms. McNeill: Right now, this area in yellow shows the boundaries. This is what Columbia looks like.

Student: That's the Columbia, South Carolina?

Student: It looks like a bird!

Ms. McNeill: See the red lines. We have these red lines in here. So… those are Interstates. You can see that this is 26. This is 26 going up this way towards Greenville. This is 77, going up towards Charlotte. And this is 20. If you go this way, you go to Florence. If you go this way, you're going to go to Aiken.

Student: Cool!

Ms. McNeill: So you can see how Columbia is. This data is what Columbia looked like in the year 2000. Look at the blue piece here. This is what Columbia looked like in 1893.

Students: Goaaaa!

Ms. McNeill: What has happened? What has happened to Columbia? Peter

Peter: Well, it grew bigger!

Ms. McNeill: Here's where we are, here is where Columbia used to be. So, what do you think happened? Did you know that it used to take…before we go there, what do you think that the people who lived out here [did]? What do you think they did for a living? If they lived in this area and this was where the city of Columbia is? What do you think they did as an occupation? Tanya.

Tanya: They used to go out and pick cotton so they could sell.

Ms. McNeill: So they were farmers, right? This was farm area. So back in 1893, what type of transportation would you have taken trying to get into the city of Columbia? Chris.

Chris: Um, by um, horse and a carriage?

Ms. McNeill: Very good. And how long do you think it would take to get from this area? Let's say you work on a farm in this area and you needed to go sell your goods in the city of Columbia, how long do you think it would take to get to the city of Columbia given that you had to take a horse and buggy? Shannon.

Shannon: A day.

Ms. McNeill: You are absolutely right! How long do you think it takes you today to get to that same spot? What type of transportation would you use today to get to downtown?

Megan: It depends on traffic.

Ms. McNeill: Right, but just in general how long do you think it takes you to get downtown?

Megan: About 15, 20 minutes.

Ms. McNeill: Exactly! It takes you about 15 to 20 minutes to get downtown. That's a big difference, isn't it?

As the whole group discussion illustrates, Ms. McNeill shared her knowledge about the interstates near Columbia, SC to expand and connect to the students' prior knowledge. She also guided the students to make their own connections and use their daily life experiences in learning geography. As Ms. McNeill's discussion and guiding questions illustrate, her prompts were directly related to the students' geography learning rather than on teaching them GIS itself.

During the small group GIS activities, Ms. McNeill monitored students' learning and directed (or re-directed) their thinking. She carefully observed and listened to the students' responses to redirect or encourage the students to think. On many occasions, her guidance made a significant difference in adding depth to the students' understanding. For example, when the students worked on project 1 (Columbia), the students were asked to explain the influencing factors for the development of their city. Even after displaying different layers of themes (rivers, lakes, railroads, and interstates) on the GIS map, only two of the six groups understood the influence of the Interstates on the growth pattern of the city.

The GIS maps that they were examining on the screen were not helpful to them because they did not have prior knowledge about the influencing factors for growth. Since they did not understand the questions about the relationship between urban growth and transportation networks, the teacher needed to guide the students' thinking, and it made the difference. What follows is an example of a conversation that the teacher had with a group that was able to identify the relationship between the city's growth and transportation.

Ms. McNeill: In which direction has it grown the most, would you say?

Megan: Right.

Tiffany: Rightwards.
Ms. McNeill: And then think of why do you think it grew that way? What do you think were some of the reasons why it grew?
Megan: The people.
Tiffany: Because… that part was up there.
Ms. McNeill: Yes, why do you think the county, the city of Columbia went this way?
Megan: This side is too crowded really.
Tiffany: Because people started…
Ms. McNeill: Because what?
Tiffany: People started coming.
Ms. McNeill: So people started coming out here, right? Why do you think, what may be some reasons why people moved out that way? Out this way? What do you think is out here that people want to move out there?
Tiffany: Their family?
Ms. McNeill: Well…
Megan: More stores?
Ms. McNeill: Maybe more… there you go… more stores. What is it? [Pointing at the screen]
Tiffany: The highway!
Ms. McNeill: Yeah.

In this illustration, the teacher displayed an overlay map of the highways on the screen and asked the students to observe it and encouraged them to find growth patterns that emerged from the map. The other groups were now focusing on people and attractions in the growing area. While Megan and Tiffany were able to make the connection between high population areas and transportation, other groups were not able to make the connection between population growth and transportation. The other groups' answers were "more people," "more stores," and "more land."

Another example is when Peter and Chris worked on the second lesson (South Carolina); the students were trying to explain why a large number of people lived in the Charleston area in 1790. When Peter and Chris focused on the size of the county, the teacher redirected the students to consider other elements in the highly populated area and guided them to think about what they previously learned during class.

Ms. McNeill: And why do you think those people live there?
Chris: It was a big county or something?
Peter: Well that looked like it's bigger than that one.
Chris: No, because it was the biggest.
Peter: It's the biggest out of all of them

Ms. McNeill: Do you think size is the only thing to do with it?
Chris: Well, why is this one empty?
Ms. McNeill: Think about what we know about 1790.
Peter & Chris: Um...
Ms. McNeill: We talked about it in class.
Peter: I know in 1790 they used boats because they couldn't go through the forest.
Ms. McNeill: So... why have the most people been on the coast?
Chris: Because it's near the water.
Peter: Because Charleston is near the water.
Chris: Yeah.
Peter: Because they used boats.
Ms. McNeill: Here you go.

Finally, after the teacher's prompt, Peter and Chris were able to use the information that they learned during the class discussion to explain why the map looked the way it did.

During the instruction, the teacher often rephrased and guided the students' responses to help the students improve their geography learning. When the students' responses were too broad, the teacher encouraged them to think about specific examples so the students could make the connection themselves between the specific examples and more general ideas. The students seemed to be highly engaged in the GIS activities while the teacher facilitated the process.

Teacher as a Constructor for the Geography Curriculum

As Ms. McNeill made instructional and curricular decisions, she served as the constructor of the classroom curriculum. She planned and directed how the content was to be introduced and the GIS technology was to be used. Sometimes she wanted to make changes because she believed that the changes would be beneficial for the students' geography learning, which was her instructional goal. Other times she needed to make some changes to accommodate time constraints and the school's schedule changes.

First, Ms. McNeill wanted to include activities that were not in the GIS module, such as adding a KWL chart and "sharing the sketch maps," into her instructional process. Before Ms. McNeill began the lesson, she indicated that she wanted to include KWL charts for each lesson. She believed that it would be beneficial to the students. Also, the sketch maps were originally included as a part of data gathering, not a part of instruction. However, Ms. McNeill indicated that the students could learn from their sketch maps if they were allowed to compare the differences between their pre and post sketch maps. Therefore, she gave the students opportunities to compare their pre and post sketch maps and to present the differences to their classmates. This approach encouraged the students to reflect on their progress. During an interview, Ms. McNeill expressed her satisfaction by saying:

> I like that we went over and let them share and talk about their maps. Because when you let them talk about their pictures, all of a sudden, you get a little more depth to it. Even though it just looked like a line, it may mean a lot more information tied to that...

Obviously, she had her reasons for the inclusion of the activity, and she was satisfied with the outcome.

Second, Ms. McNeill made some changes in terms of the content of classroom discussion. It occurred especially when the students did not seem to have enough prior knowledge or did not seem to make connections to their prior knowledge as she had anticipated. For example, when the students did not have enough prior knowledge about their state, the teacher extended the discussion period and started with what the students knew in order to guide them toward the main focus of the lesson. Ms. McNeill explained the students' lack of prior knowledge by saying "Several of them know Myrtle Beach. Otherwise they have no concepts of the rest of South Carolina. I'm trying to think of... somewhat help generate some sort of idea of what the rest of state looks like." In contrast, on a few occasions, Ms. McNeill decided to stop the activity if the students were not engaged. For example, while Ms. McNeill was sharing a picture book that showed an African-American community and people in the early 20th century, the students became excited, made loud noises, and meaningless comments. Ms. McNeill closed the book and said to the students, "we stop here. If you are not paying attention, we are wasting our time." Later during the interview, she explained "As soon as it happens, then we just stopped because it's not going anywhere. They weren't going to get anything out, rather than laughing."

Third, the teacher reorganized the unit as she encountered issues along the way. During the fifth session, for example, the instructional process did not go well. Ms. McNeill reflected on the lesson saying "too much activities and too much content at the same time," and the students were experiencing an "overload." After the class, she reorganized the lessons to include fewer activities and to provide the students with more specific guidelines. Another example is that she combined lesson 3 (the United States) and lesson 4 (Inquiry project) since the previous lessons took longer than Ms. McNeill had anticipated and she wanted to give the students sufficient time for their inquiry project.

Clearly, all of her decisions were directly related to the students' geography learning, rather than learning about GIS. Whenever Ms. McNeill made changes, she did it to provide more opportunities for the students to reflect on what they were learning, to provide more discussion to expand their prior knowledge, and to provide more time to work on their project.

Implications

Throughout the instructional process, the teacher guided the students' learning and made curricular and instructional decisions. The teacher answered the students' questions and carefully monitored the students' progress to expand their geography learning. There are several implications of these findings for teachers' GIS training.

This study suggests that teachers should not only know about GIS technology, but also should be able to make numerous instructional decisions that are beneficial to the students' learning. The key to the success of the instruction was not how much the teacher knew about GIS, but instead how the teacher guided the students and facilitated their learning. If the implementation of GIS in elementary classrooms is to be successful, teachers should be able to monitor students' progress or lack of progress, should guide them to use their prior knowledge, and should be able to direct (or re-direct) their geographic thinking paths.

When GIS experts provide workshops for teachers, the priority has been on teachers' learning to use GIS software. The remainder of the implementation process, such as when to use GIS within the existing curricula, how to introduce it, and what kinds of discussion questions could be asked, has been left for teachers to wrestle with in isolation. Because the teacher in this study was highly skilled, she was able to apply her teaching skills in the integration of GIS. Given the importance of teacher decision-making in the process of GIS infusion, geography educators should consider providing pedagogical guidance that would be useful for teachers learning "how to teach," in addition to "what to teach" with GIS.

In this study, the teacher spent a great deal of time and effort to make connections to the students' prior knowledge and their daily experiences. It was challenging when the students did not have enough prior knowledge. In these cases, the teacher had to go back and start from what the students knew. Although the teacher expected some level of challenge, it was the major reason that the teacher had to spend more time on her instruction than she planned, and it caused another time constraint problem in the subsequent lessons.

While most of the published GIS curriculum materials extensively explain the GIS functions, relatively small portions are usually devoted to enhancing students' background knowledge. The notion that teachers need to know GIS to teach GIS is understandable. Yet, in developing GIS materials, geography educators need to pay more attention to providing resources that could help students expand their background knowledge, since the success of the lesson is not about GIS itself, but about how well students learn about geography.

Ms. McNeill showed clearly that her goal was not teaching about GIS, but using GIS as a means to teach geography. She led discussions to encourage the students to think geographically and made changes that would be beneficial to the students' geography learning. Her lack of confidence in using GIS technology was actually not a challenging issue in teaching geography using GIS. Rather, the success of the lesson seemed to be based on her clear instructional goal: the advancement of students' geography learning.

RECOMMENDATIONS

When teachers implement GIS in their classrooms, they need to make numerous decisions regarding guiding students' geography learning. They monitor students' responses and direct their thinking. However, most of the GIS training opportunities and materials developed do not include how to help teachers' pedagogical decision making and/or increase students' background knowledge. The following recommendations focus on how to better prepare in-service and pre-service teachers to integrate GIS technology in a social studies curriculum.

Recommendations for Teacher Educators

When providing GIS training or developing GIS materials:

- Include examples of instructional strategies, such as appropriate teaching strategies and group activities
- Share guiding questions that are related to the specific spatial thinking addressed with the content of the GIS training.
- Inform teachers about what kinds of background knowledge are essential for the specific content to be taught with GIS
- Include resources that can enhance students' background knowledge, such as children's literature, web addresses, vocabulary lists, etc.

Recommendations for Classroom Teachers

When implementing GIS in your classroom:

- Focus on your students' geography learning, instead of teaching about GIS.
- Think though how you could guide students' learning of geography by preparing possible guiding questions that would be appropriate to your students. You know your students the best.
- Monitor often for students' understanding by asking questions and listening for misconceptions.
- Be ready to make changes and modify the GIS materials. Remember, no materials are specifically developed for you, other than your own.

NOTES

1. In order to protect the research participants' confidentiality, pseudonyms were used for the school, the teachers, and the students.
2. PACT is a standardized test that students take in South Carolina.
3. The PACT scores from six students who did not enrolled in Chantilly Elementary School in their second grade were not available.
4. Just as the PACT English score, the same six students' mathematics test results were not available.

REFERENCES

Alibrandi, M. (2003). *GIS in the classroom: Using geographic information systems in social studies and environmental science.* Portsmouth, NH: Heinemann.

Alibrandi, M., & Sarnoff H. (2006). Using GIS to answer the 'whys' of 'where' in social studies. *Social Education, 70,* 138–143.

Baker, T. R., & White, S. H. (2003). The effective of GIS on students' attitudes, self-efficacy, and achievement in middle school science classrooms. *Journal of Geography, 102*(6), 243–254.

Bednarz, S. W. (2004). Geographic information systems: A tool to support geography and environmental education? *GeoJournal, 60,* 191–199.

Blissett, G., & Atkins, M. (1993). Are they thinking? Are they learning? A study of the use of interactive video. *Computers & Education, 21*(1–2), 31–39.

Budin, H. R. (1991). Technology and the teacher's role. *Computer in the Schools, 8*(1–3), 15–26.

Clandinin, D. J., & Connelly, F. M. (1992). Teacher as curriculum maker. In P. W. Jackson (Ed.), *Handbook of Research on Curriculum* (pp. 363–401). New York: Simon & Schuster Macmillan.

Creswell, J. W. (1994). *Research design: Qualitative & quantitative approaches.* California: SAGE Publication.

Doolittle, P. E., & Hicks, D. (2003). Constructivism as a theoretical foundation for the use of technology in social studies. *Theory and Research in Social Education, 31*(1), 72–104.

ESRI. (2000). *GIS in schools.* Redlands, CA: Environment Systems Research Institute, Inc.

Frenkel, J. R., & Wallen, N. E. (1996). *How to design and evaluate research in education.* New York: McGraw-Hill, Inc.

Gatrell, J. D. (2004). Making room: Integrating geo-technologies into teacher education. *Journal of Geography, 103*(5), 193–198.

Geography Education Standards Project. (1994). *Geography for life: National geography standards.* Washington DC: National Geographic Society.

Glaser, B. G., & Strauss, A. L. (1967). *The discovery of grounded theory.* Chicago: Aldine.

Glesne, C. (1999). *Becoming qualitative researchers: An introduction.* New York: Addison Wesley Longmann.

Griffin, G. A. (1994). Teacher education curriculum in a time of school reform. In M. J. O'Hair and S. J. Odell (Eds.), *Partnerships in education: Teacher education yearbook II* (pp. 224–245). Fort Worth: Harcourt.

Keiper, A. T. (1999). GIS for elementary students: An inquiry into a new approach to learning geography. *Journal of Geography, 98*(2), 47–59.

Kerski, J. J. (2001). A national assessment of GIS in American high schools. *International Research in Geographical and Environmental Education, 10,* 72–84.

Kerski, J. J. (2003). The implementation and effectiveness of Geographic Information Systems technology and methods in secondary education. *Journal of Geography, 102*(3), 128–137.

Koo, R., & Harlin, R. P. (2001). Learning in nontraditional environments: An international perspective. *Childhood Education, 77*(6), 338–340.

Lincoln, Y. S., & Guba, E. G. (1985). *Naturalistic inquiry.* Thousand Oaks, CA: Sage.

Mason, C. L., & Alibrandi, M. (1999). Teaching geography. In J. A. Braun, Jr. & C. F. Risinger (Eds.), *Surfing Social Studies* (pp. 41–48). Washington D. C.: National Council for the Social Studies.

Merriam, S. B. (1998). *Qualitative research and case study applications in education.* San Francisco: Jossey Bass.

Milman, N. B., & Heinecke, W. F. (2000). Innovative integration of technology in an undergraduate history course. *Theory and Research in Social Education, 28*(4), 245–565.

Milson, A. J. (2002). The Internet and inquiry learning: The integration of medium and method in a sixth grade social studies classroom. *Theory and Research in Social Education, 30*(3) 330–353.

Molebash, P. E. (2002). Constructivism meets technology integration: The CUFA technology guidelines in an elementary social studies methods course. *Theory and Research in Social Education, 30*(3), 429–455.

Nolan. J., Jr., & Meister, D. G. (2000). *Teachers and educational change: The lived experience of secondary school restructuring.* Albany, NY: State University of New York Press.

Olson, J, K., & Clough, M. P. (2001). Technology's tendency to undermine serious study: A cautionary note. *The Clearing House, 75*(1), 8–13.

Rasku-Puttonen, H., Etelapelto, A., Arvaja, M., & Hakkinen, P. (2003). Is successful scaffolding an illusion?—Shifting patterns of responsibility and control in teacher-student interaction during a long term learning project. *Instructional Science, 31*(6), 377–393.

Rose, A. S., & Winterfeldt, F. H. (1998). Waking the sleeping giant: A learning community in social studies methods and technology, *Social Education, 62*(3), 151–52.

Shin, E. (2006). Using Geographic Information System (GIS) to improve fourth graders' geographic content knowledge and map skills. *Journal of Geography, 105*(3), 109–120.

Shulman, L. S. (1986). Those who understand: Knowledge growth in teaching. *Educational Researcher, 15*(2), 4–15.

Shulman, L. S. (1987). Knowledge and teaching: Foundations of the new reform. *Harvard Educational Review, 57*(1), 1–22.

Soltis, J. (1990). The ethics of qualitative research. In E. W. Eisner & A. Pershkin (Eds.), *Qualitative inquiry in education: The continuing debate* (pp. 247–257). New York: Teachers College Press.

South Carolina Department of Education. (2003). *School reportcard.* Retrieved on December 16, 2006, from http://ed.sc.gov/topics/researchandstats/schoolreportcard/2002/default.cfm

Spradley, J. P. (1980). *Participant observation.* Orlando, FL: Harcourt Brace & Company.

Thornton, S. J. (1991). Teacher as curricular-instructional gatekeeper in social studies. In J. P. Shaver (Ed.), *Handbook of Research on social studies teaching and learning* (pp. 237–248). New York: Macmillan.

Tiene, D., & Luft, P. (2001). Classroom dynamics in a technology-rich learning environment. *Learning and Leading with Technology, 29*(4), 10–13, 60.

Wang, Y. (2001). Student teachers' perception and practice of the teachers' role when teaching with computers. *Journal of Educational Computing Research, 24*(4), 419–434.

West, B. A. (2003). Student attitudes and the impact of GIS on thinking skills and motivation. *Journal of Geography, 102*(6), 267–274.

Wiegand, P. (2003). School students' understanding of choropleth maps: Evidence from collaborative mapmaking using GIS. *Journal of Geography, 102*(6), 234–242.

Wigglesworth, J. C. (2003). What is the best route? Route-finding strategies of middle school students using GIS. *Journal of Geography, 102*(6), 282–291.

Wilder, A., Brinkerhoff, J. D., & Higgins, T. M. (2003). Geographic Information Technologies + project based sciences: A contextualized professional development approach. *Journal of Geography, 102*(6), 255–266.

Wiske, M. S. (Ed.). (1998). *Teaching for understanding: Linking research with practice.* San Francisco: Jossey-Bass, Inc.

CHAPTER 15

DIVING IN

Preparing Teachers to Implement Geospatial Technologies in the K–12 Classroom

Kristin J. Alvarez
Keene State College

The Geographic Alliance network began in 1986 as part of an effort by National Geographic Society to build grassroots organizations of classroom teachers and academic geographers joined together in an effort to promote and support geographic literacy in the K–12 classroom. Each state was encouraged to establish such an alliance and the National Geographic Society provided much of the original funding. One of the primary concerns in establishing these alliances was the fact that many classroom teachers were not trained in geography and "teachers teach what they know" (Douglas, 1999). State Geographic Alliances were charged with the task of first teaching the teachers the geographic content; those teachers would, in turn, teach their peers and their students. This system was couched in the goal of empowering teachers—empowered to be an instructional leader among their peers and to be an effective teacher of geographic content to whatever grade level they taught.

The New Hampshire Geographic Alliance (NHGA) grasped this charge from the National Geographic Society and joined with the New Hampshire Department of Education in the late 1980s. The NHGA has been an

Digital Geography: Geospatial Technologies in the Social Studies Classroom, pages 291–302
Copyright © 2008 by Information Age Publishing
All rights of reproduction in any form reserved.

active player in geographic education, as well as social studies education and professional development for teachers across the state. Using the model provided by the National Geographic Society, early summer institutes were generally focused on basic geography content and its associated pedagogy. The one or two week institutes were packed with geographic content and centered on preparing teachers to return to their school settings armed with geography content, and an enthusiasm for that geography content that would drive them to share that enthusiasm with others, both peers and students.

A cadre of teachers across the state who participated in those early institutes continued to be hungry for more geographic knowledge and associated pedagogy. They also encouraged their peers to be involved in this enlivenment of geographic education. This demand caused those involved with NHGA to search for an array of topics for workshops and summer institutes that would pique the interest of those already deeply involved in geographic education, as well those who were simply interested in learning more geography. The result was a variety of workshops on different topics and summer institutes with such themes as the old textile mills of New Hampshire and the significance of stone walls. It is only natural that the impact of geospatial technology on the global society would become a source of curiosity and excitement in the world of K–12 classrooms, and thus become a focus of workshops and summer institutes as part of the NHGA program to enhance geographic education.

The significance of Geographic Information Systems (GIS) on the world of spatial analysis is without question. Susan Hanson, former president of the Association of American Geographers, included GIS as one of the ten geographic ideas that changed the world in her edited book by the same name (1997). GIS is possibly the most visible of several geospatial technologies (GST) that have become a part of our everyday lives, but there are other types of geospatial technologies that impact us as well. In terms of potential use in the K–12 classroom, the term *geospatial technologies* refers to GIS, global positioning system (GPS), remote sensing, and web-based map viewers.

After several workshops utilizing some aspects of GST for the K–12 classroom in New Hampshire, it was time for a full-fledged summer institute combining several GST and their classroom application. Thus was born the 2006 NHGA Summer Institute—*GIS and Geospatial Technologies in the Classroom.* This institute was designed for New Hampshire teachers to learn how to incorporate these technologies, not only in their geography content, but within all the social studies areas. These technologies have broad application both in intra-disciplinary use within the social studies and interdisciplinary application across the curriculum. What follows is a case study of this NHGA summer institute.

THE SUMMER INSTITUTE

Structure

On July 9, 2006, thirteen teachers gathered on a small liberal arts college campus to begin a six-day residential summer institute to learn how to engage their students in new and powerful ways. These teachers applied for the summer institute after having read about it in the NHGA newsletter, the NHGA listserv, or hearing by word of mouth from their peers. The cost to the teachers for the institute was $175 which included most meals, housing on campus, books and software, field trips, and instruction. Additionally, graduate credit in geography was available for $113 per credit. This graduate credit is most commonly used to complete teacher re-certification required every three years in New Hampshire.

Sessions would be held primarily in the GIS lab in the Geography Department, but also in the dorm area, some classrooms, and in the field. It was relatively simple for the institute planners/leaders to determine specifically which GST to include in this short time. It was far more difficult to decide what the sequence of content would be. Should we start them off with the high interest, relatively simple web-based map viewers to grab their interest? Or would it be more beneficial to begin with the meat of geospatial technologies, GIS? The institute leader with extensive middle school classroom expertise in these technologies urged that we begin with the most difficult task of working directly with GIS software. With serious reservations concerning this decision, the agenda was set.

Participants

The participants were from across the state of New Hampshire. They ranged from teachers with more than 25 years of experience to a senior pre-service elementary education/geography major from Keene State College where the institute was held. Some had previous NHGA experiences through workshops and institutes and others had none. The group included 4 males and 9 females. Two of the males and one of the females commuted to their homes each night and did not stay in the dormitories. The grades taught by participants ranged from 3rd grade through high school (see Table 15.1). Six participants taught social studies subjects. There was one middle school science teacher, one high school media specialist, and one 7–12 technology educator. Three participants were elementary teachers who taught all subjects and one participant was a pre-service teacher preparing to begin student teaching at the elementary level. Previous GST experience ranged from none at all to a smattering of exposure to

TABLE 15.1
Summer Institute Participants

Teacher	Grade(s) taught	Subject(s) taught	Previous Geospatial experience
Patsy	7	social studies, math	none
Nancy	9–12	media specialist	2 NHGA GPS workshops
Ron	7–12	technology education	some GPS
Ann L.	6,7,8	geography, U.S. history, language arts, religion	little GPS, *Google Earth*
Murray	9–12	social studies, archaeology	some exposure to GIS, GPS
David	4	all—elementary	extensive GPS
Melissa	7,8	social studies, geography	GIS course 10 years ago, uses a little in classroom with GPS
Michelle	8	social studies, health	none
Ann R.	6	science	GPS workshop, *Google Earth*
Geralynn	3–4	all—elementary	extensive GIS, some GPS
Carol	7	U.S. history	2 NHGA GPS workshops
Sean	4	all—elementary	2 NHGA GPS workshops
Jenna	pre-service	all—elementary	GIS course

one teacher (elementary) who had graduated from college in the late 1990s with a degree in elementary education and geography, taught elementary school for a short time, left to work as a GIS analyst for several years, and then returned to the elementary classroom as a grades 3–4 looping teacher.

Facilitators

Four people comprised the facilitator staff for the NHGA Summer Institute: the two NHGA co-coordinators, a business partner of the Alliance who is the founder of a large spatial technology company in New Hampshire, and a person from the region who is well-known nationally in GIS in education. There were also three guest speakers who came at various times during the institute.

Institute Activities

For ease of understanding the focus and flow of the experience, this section is divided into each day of the institute.

Sunday

Institute participants arrived late in the afternoon, settled into their lodgings, and gathered for pizza, introductions, and an activity. Since the focus of the activity was technology, the evening's program examined some earlier spatial technology such as the sextant and astrolabe. There was opportunity for hands-on practice with the sextant. *Longitude* by Dava Sobel (1995) was distributed to all participants to read before viewing the video version of the book the following evening. A spirited discussion concerning geospatial technology ensued.

Monday

The first morning session provided an overview of remote sensing and an exercise for students of many grade levels was modeled. The exercise was adapted from the Wetlands Education Through Maps and Aerial Photography (WETMAAP) materials (http://www.wetmaap.org). This was followed by an overview of GIS. Both GIS texts provided for institute participants were distributed and reviewed (Gorr & Kurland, 2005; Malone, Palmer, Voigt, Napoleon, & Feaster, 2005). The GIS software used for this portion of the institute was *ArcGIS 9.1.* The participants were encouraged to choose one of the texts and move at their own pace through either several modules or tutorials, depending upon the text. Facilitators circulated to assist participants. As this was an intensive day, participants were encouraged to take frequent breaks. The evening program was the videotape version of *Longitude.* This contrast to the GIS technology of earlier in the day and the comparison with the frustration felt by both the participants and the subject of the book/videotape, Harrison, served to be an apt activity for ending the day.

Tuesday

Participants continued using the self-paced work on both of the GIS texts in the GIS lab. Late in the afternoon, the pace changed with an overview of GPS technology. The participants were all given GPS units to use to practice finding coordinates. This was followed by an overview of geocaching; the leisure time activity using GPS to locate hidden caches posted on the Internet (see http://www.geocaching.com/).

Wednesday

This day was spent on a field trip with multiple destinations. Each van carrying participants was equipped with a navigation system so that wayfinding could be aided by the use of a GPS. The participants were given printouts of a geocache to be found on the way toward our first destination. The first stop was in a town forest approximately 15 miles from the campus where the participants used the GPS skills presented the day before to locate a published geocache. Success was at hand within just a few minutes.

The next stop on the field day was at TeleAtlas, a corporation in Lebanon, NH that produces digital maps and content for navigation and location-based services (LBS). The participants were able to observe in small groups the GIS technicians and analysts at work and interact with them. A presentation for the participants introduced them to what they would be observing and the role of GIS and GPS in this industry. Later in the day the group traveled to the last stop on our field day, the United States Army Corps of Engineers' Cold Regions Research and Engineering Laboratory (CRREL) also located in Hanover, NH. This facility is used to test materials and equipment used in arctic conditions. The facility also houses the GIS Center for the Army Corps of Engineers and the participants were given a tour of the GIS Center and a presentation on the work done there. Of particular interest in this presentation were the Hurricane Katrina flood maps done by the Corps at this location.

Thursday

The morning began with guided practice using *ArcExplorer: Java Edition for Education* (*AEJEE*), a free downloadable GIS software product (http://www.esri.com). This was followed by more guided practice using another software product designed for the K–12 classroom—*My World GIS* (see http://www.pasco.com/myworld/about.html). The focus of the day moved from alternative classroom GIS software such as *AEJEE* and *My World* to web-based map viewers. The first map viewer introduced was *Google Earth*, commonly called a digital globe with GIS-like layers (see http://www.earth.google.com). After exploring this geospatial tool, the participants were introduced to *SketchUp*, a powerful 3-D drawing program that may be used in conjunction with *Google Earth* or separately (http://www.sketchup.com). Another web-based map viewer explored by participants was *TerraServer-USA* (http://www.terraserver-usa.com). This website allows the user to view either topographical maps or aerial photographs of almost any location in the country. These maps are downloadable and may be used in conjunction with ArcGIS. The evening program concluded with a presentation from the NH representative for *My Wonderful World*, a collaborative effort to enhance geographic education among several groups including the National Geographic Society, the Association of American Geographers, the National Council for Geographic Education and several other public and private organizations (http://www.mywonderfulworld.com).

Friday

The closing day of the institute began with a presentation from a GIS specialist from the NH state GIS clearinghouse, *NH Granit* (http://www.granit.sr.unh.edu). She introduced the participants to this data source for use in their classrooms and also demonstrated the web-based GIS mapping tool on their website. This was followed by a presentation from a representative of the NH State Archives and Record Management (http://www.sos.nh.gov/archives). This presenter was a long-time surveyor in New Hampshire until he became interested in archiving data and maps for future generations. He shared extensively about the historical maps available for free at the State Archives. After debriefing and evaluation, the summer institute officially ended with plans to have follow-up sessions in order to maintain the enthusiasm and enhance the geospatial technology skills that had been introduced.

Summary of Institute Activities

The thirteen participants began by working with remote sensing and moved quickly into ArcGIS for two days. This was followed by an introduction to GPS, a geocaching experience, and a field trip to two geospatial facilities. The participants then worked with alternative GIS software designed specifically for the K–12 classroom and explored some web-based map viewers and web-based GIS applications. The institute closed with presentations about data available to classroom teachers in New Hampshire. In short, the participants moved from very skill-based ArcGIS to the more user-friendly web-based GST. Embedded in all activities was pedagogical practice appropriate to GST. For example, guided practice was used in introductory GPS activities and culminated in a real-world geocaching experience. Remotely sensed images were analyzed and ground truthing photos were used in order to provide self-assessment of their own analyses. After each type of technology was introduced the participants were asked to share what ideas they had for applying the technology in their own school settings and real-world, inquiry-based applications were always encouraged.

Follow-Up Activity

The participants were invited to a follow-up activity held in the fall following the summer institute at the NH State Archives. The archivist who presented at the summer institute opened the State Archives for those who were able to attend. The teachers downloaded historical maps after exploring some of the data available to them. The group then adjourned to the classroom of one of the institute facilitators who demonstrated how to apply digital images to GIS maps. An unintended outcome of the follow-up activity was that an active NH Geographic Alliance member (a middle school social studies teacher) who did not attend the summer institute but who had some rudimentary knowledge of GST, asked if she could join in for this follow-up activity. She has joined this cohort group in implementing GST into her classroom.

ANALYSIS

Initial evaluations completed by the participants were most positive. All indicated that the sessions were clear and of high interest. Their evaluations also confirmed that the facilitators had chosen the correct sequence of content; moving from most difficult and challenging to easy and fun. The facilitators felt that having some GIS experience in the beginning would provide a foundation for the possible implementation of other geospatial activities. For example, it may be fun to have students use a GPS on the school campus, but if the teacher understands how GPS can be used along with GIS, he or she can build authentic inquiry-based lessons for students that will provide them the skills to carry into their future endeavors. Additionally, the more difficult tasks at the beginning of an institute when teachers are fresh may be more effective. The evaluations indicated participants' enthusiasm for implementing the GST in their classrooms. Several were realistic in their assessment of the realities of the constraints of time and money in their particular classrooms.

The true measure of the effectiveness of the institute is whether or not the teacher actually implements GST in the classroom. It is difficult to assess the effectiveness of the institute within a specific timeframe. When a teacher participates in a professional development experience, that learning may be quickly applied in the teacher's classroom or ideas and plans may ruminate in the teacher's mind for some time before they are implemented.

In the relatively short time since the institute, there has been evidence of the adoption of the tools introduced by some of the participants. One wonderful example was demonstrated by David's fourth grade class when they produced a GIS map (Figure 15.1) of a conservation area adjacent to their elementary school. This class did extensive GPS work in compiling the conservation land map and has done geocaching as a class.

Another participant, Patsy, who teaches social studies and math in 7th grade, has formed an after school Geography Club in her school and they are focusing on GST activities. Geralynn, an elementary teacher, has incorporated many, many aspects of the skills and resources from the summer institute into her classroom. She has truly integrated these skills across the

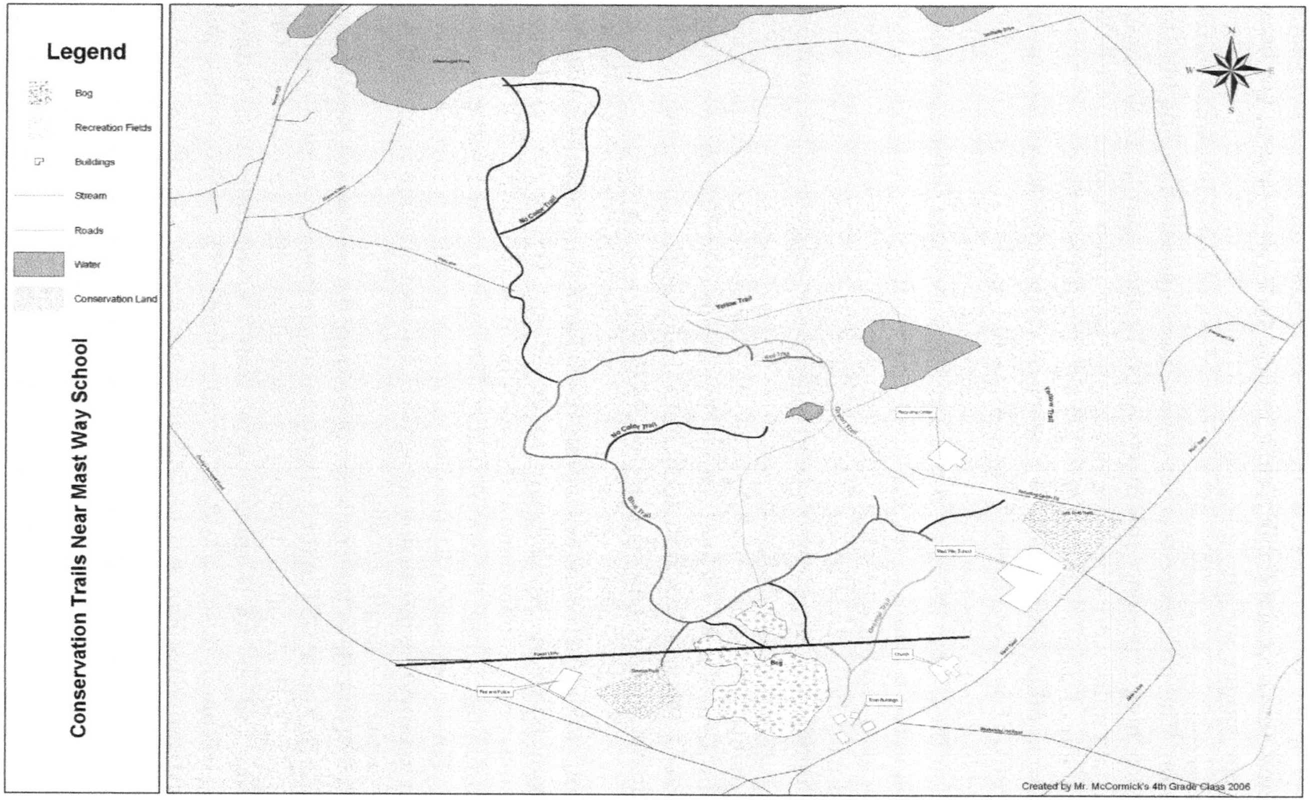

Figure 15.1. David's 4th grade class map of adjacent conservation area.

curriculum as evidenced by the example of her "inspiration journals." The students are required to find a special location outside and draw an illustration of any subject they choose to observe. They then write everything they are thinking and feeling while making their observations. On occasion, she has the students go outside with their GPS units and their "inspiration journals" where they mark a waypoint and log the journal entry's location. The pre-service teacher, Jenna, has commented during her student teaching experience this year that she thought many times about how she could utilize GST when investigating regions and is hoping to have the availability of these technologies in her own future classroom.

Several other participants have reported varying degrees of incorporation of GST in their classes. Some have commented that as much as they would like to incorporate these technologies, time and money constraints prevent them from doing so at this time. One example of the constraints encountered is found in the experience of two institute participants, one elementary teacher and one middle school teacher, who teach in the same small parochial school in an economically depressed region of the state. The middle school teacher is faced with at least 5 subject preparations weekly ranging from language arts to social studies to religion. The time constraints are further exacerbated by the budget restraints experienced by this particular K–8 school inasmuch as funding for hardware and software for GST must be searched out by the individual teachers and they simply do not have the time to make that extra effort.

It is clear that barriers exist in the implementation of the skills and information garnered in the institute. Such barriers include the availability of both hardware and software to implement GST in the classroom, demands on classroom time, and development of standards-based learning opportunities across the social studies (and perhaps across the curriculum). Finally, the major barrier lies within the teacher. Is there a commitment to providing instruction using GST? This last question is certainly not an indictment—simply a reality.

The NHGA views the challenge of helping the teachers of New Hampshire incorporate GST in their classrooms as a primary concern. To address this need, NHGA will be holding two workshops this year—one for GPS training and one for learning how to create a Community Atlas (http://www.esri.com/industries/k-12/atlas/index.html). The Alliance has purchased two classroom sets of GPS units for teachers to borrow free of charge. These units have been in high demand. The facilitators of the summer institute quickly realized that follow-up sessions are very important in maintaining the momentum of the summer institute. These follow-up sessions focusing on both skills and pedagogy should be set at the conclusion of the institute so there can be a level of commitment among those teachers who would like these opportunities. By waiting until later to plan these sessions some of the momentum and accountability is lost.

Another avenue available to meet the challenge is to use the Alliance's communication tools such as the listserv, the newsletter, and the website to promote best practices in GST. This can possibly be accomplished through the publishing of teacher-tested lessons. Another approach would be to sponsor a day-long workshop where teachers bring ideas to share in the classroom implementation of GST. Since ESRI, the producer of ArcGIS, hosts an annual Education Users Conference, it may be beneficial to offer funding for one or two teachers to attend that conference with the proviso that they present a GST workshop in their district or region the following school year. This opportunity to attend the conference could be competitive and may require lessons to be published.

CONCLUSION

Three important realities exist in education today. The first is that social studies have been left behind in the rush to accountability. As a result, social studies instruction has been greatly diminished at the elementary level. The impact of that diminishment will soon be felt at the secondary levels. The second reality is that GST are one of the fastest growing industries in the United States according to the US Department of Labor (2004) and our students need to be prepared to meet the demands of this growing industry. Recent research into spatial thinking (National Research Council, 2006) has reported that it is important to integrate spatial thinking across the curriculum. It is believed that GIS has an important role to play in building spatial thinking among students. These three realities combine to present a clear agenda for the NHGA and other groups with a focus on geographic education. That agenda must include practical and meaningful training for teachers in GST to meet the challenge of these three realities.

REFERENCES

Douglas, M. P. (1999). *The history, psychology, and pedagogy of geographic literacy.* Westport, CT: Praeger Publishers.

Gorr, W., & Kurland, K. (2005). *GIS tutorial.* Redlands, CA: ESRI Press.

Hanson, S. (Ed.). (1997). *Ten geographic ideas that changed the world.* New Brunswick, NJ: Rutgers University Press.

Malone, L., Palmer, A., Voigt, C., Napoleon, E., & Feaster, L. (2005). *Mapping our world: GIS lessons for educators.* Redlands, CA: ESRI Press.

National Research Council. (2005). *Learning to think spatially.* Washington, DC: The National Academies Press.

Sobel. D. (1995). *Longitude: The true story of a lone genius who solved the scientific problem of his time.* New York: Penguin Books.
United States Department of Labor. (2004). *The president's high growth job training initiative.* Retrieved December 29, 2006, from http://www.doleta.gov/BRG/JobTrainingInitiative

CONCLUSION

MAPPING GEOSPATIAL TECHNOLOGIES IN SOCIAL STUDIES EDUCATION

Andrew J. Milson
University of North Texas

The expanse of the geospatial technology landscape and the varieties of classroom and research expeditions upon which one might embark with geospatial technologies (GST) in social studies education cannot be fully captured in one volume. The reality of technology in our world today is that the terrain is always expanding and shifting with new developments and opportunities. Our hope, though, is that the work of the authors that is presented in this book will inspire further mapping and exploration of this exciting terrain. Through this book, we have attempted to survey the shifting sands of GST in social studies education and to map some of the contours that exist in this field in 2007. There are several lines that emerge as significant to the orientation of GST in education during this first decade of the 21st century. At the risk of overextending my cartographic surveying metaphor, I would like to conclude the book by discussing a few of the benchmarks that have been identified by the authors of the chapters in this volume.

As I read the chapters of this book, the Beatles tune "With a Little Help from My Friends" rang through my head. When working with GST, I have sometimes felt a bit like Ringo Starr singing, "What would you do if I got

Digital Geography: Geospatial Technologies in the Social Studies Classroom, pages 303–306
Copyright © 2008 by Information Age Publishing
All rights of reproduction in any form reserved.

lost in ArcView?" Joseph Kerski assures us all that the International Geospatial Education Community is a growing network ready to help each of us "get by." In fact, Kerski explains in Chapter 3 that the very nature of these tools requires collaboration: "The reason why the community developed is rooted in the fact that the implementation of a new set of tools and methods in any educational environment is highly dependent on the interaction of people" (Kerski, this volume).

Indeed, the importance of collaboration and mutual support is evident in the work of each of the authors. Marsha Alibrandi and Tom Baker make it clear in their social history and timeline that networks of individuals supporting each other have been a vital component of the past two decades of work in this field. This is evident in the work of Sven Fuhrmann, Alan MacEachren, Mark Gahegan, and Roger Downs as they demonstrate the importance of collaboration among a variety of stakeholders during the development phase of a geospatial technology application bound for classroom use. Furthermore, at the implementation stage, each of the chapters in the *Social Studies Expeditions* section of the book emphasizes the importance of collaboration within the K–12 classroom when using GST. Danny Edelson, Dave Smith and Matthew Brown make a compelling argument that students must engage in collaborative analysis of geographic data if the promise of GST in educational environments is to be realized. Similarly, Aaron Doering, George Veletsianos, and Cassie Scharber remind us that GST are constructivist tools and that research on their implementation should be guided by appropriate pedagogical, theoretical, and methodological frameworks. Likewise, a collaborative multi-disciplinary framework for research is proposed by Rick Bunch, Elisabeth Nelson, Bob Lloyd, Michael Kane, and Thomas Tricot in their chapter outlining an Instructional Geographic Information Science framework for GST in education. In the final section of the book, Sarah and Bob Bednarz, Eui-kyung Shin, and Kristi Alvarez each articulate the importance of collaborating with teachers to help them prepare for—as well as to support them along—their journeys with GST in their classrooms. The significance of collaborative efforts among the International Geospatial Education Community is apparent in our history, our technical and curriculum development efforts, our classroom practice, our research, and—perhaps most importantly—our teaching of K–12 students.

A related benchmark that appears significant in our efforts to map our field is the importance of collapsing barriers and making connections. David Green and Joanna Mouatt provide us with an excellent example of using GST—in this case Google Earth—to collapse the boundaries of the school walls and the barriers associated with field-based work. They offer an avenue for establishing stronger connections with the physical environment through GST. Joseph Kerski describes how the connectivity inherent in the latest Internet-based GIS environments offers opportunities for more students and teachers to use GIS in the classroom, and thus put the world at the students' fingertips. Likewise, the authors of each of the chapters in the *Theory and Research* section argue for the importance of establishing connections to GST in educational research and practice by overcoming barriers. Jay Gatrell and John Conant suggest that GST can serve in supporting an integrative conception of the social studies curriculum that collapses the perceived boundaries between the disciplines, and thus globalizes social studies education in order to connect students with the 'flat world' of the 21st century. Shannon White demonstrates how phenomenographic research can illuminate the conceptions of GIS held by teachers and students, and how these conceptions can influence the rate at which this innovation is diffused. Similarly, Aaron Doering and his colleagues, as well as Rick Bunch and his colleagues, argue in favor of boundary-spanning research efforts that connect multiple disciplinary and theoretical traditions in an effort to improve the use of GST in education.

A third idea that emerges from the chapters of this volume is the importance of compatibility. In order to collaborate and connect, we must be compatible. At its core, compatibility requires mutual understanding and the ability to communicate effectively. Several authors in this volume have emphasized the significance of compatibility among teachers, researchers, theorists, software developers, policy makers, and teacher educators if GST is to be a successful innovation in K–12 social studies education. The *MapStats for Kids* (Fuhrmann, et al.), *My World GIS* (Edelson, et al.), and *GIS for History* (Radinsky) projects all demonstrate the importance of compatibility of the technology with the intended environment. In this case, the intended environment happens to be the very multifaceted world of K–12 education. The realities of this multifaceted environment cannot be ignored in the quest to "plug in" GST. In our chapter, Jennifer Roberts and I explore how the state standards for geography education might serve to either impede or facilitate greater compatibility between GST and high school geography teachers and students. Similarly, Sarah and Bob Bednarz discuss how greater attention to spatial thinking and expert support structures during in-service teacher education can aid in more effective implementation—and thus compatibility—of GST in education. Eui-kyung Shin's research also points out the vital role of the classroom teacher in determining how compatible GST will be with the education of elementary school students. Additionally, Shannon White's dissertation research has applied Rogers' diffusion of innovations theory to the uptake—or compatibility—of GIS in schools and explored how teachers' and students' conceptions are critical to this process.

The notions of collaboration, connection, and compatibility are certainly not new to the educational landscape. Philosophers of education have reminded us of these virtues repeatedly, yet historians of education have documented the litter of educational trends whose proponents ignored the realities of the school and the larger social environment. I believe that the authors who have contributed their work to this volume have echoed the calls for greater collaboration, connection, and compatibility. In the process, they have issued a challenge to the International Geospatial Education Community to keep these goals firmly in mind so that geospatial technologies in social studies education are not ultimately consigned to the dustbin of educational fads. There is still a great deal of terra incognita to be mapped in this field. Marsha and I hope that the readers of this volume will join the community and that the work of these authors will inspire further expeditions with GST in social studies classrooms around the globe.

ABOUT THE CONTRIBUTORS

Marsha Alibrandi is Assistant Professor of Secondary Education at Fairfield University in Connecticut, and author of *GIS in the Classroom* (2003). She taught 9–12 geography and social studies at Cape Cod Tech, and has taught undergraduate and graduate courses in geography and social studies education since 1990. While associate professor at North Carolina State University, she offered a GIS in Education course between 2000 and 2005. She continues to seek, support, research and report on classroom applications of GST nationally and internationally, and is interested in comparing spatial and linguistic cognition, and the critical, ethical and equity issues inherent in geospatial technologies. Her publications have been targeted on integrating GST in Social Studies education. She may be reached at marsh@cape.com.

Kristin J. Alvarez is an Associate Professor of geography and education at Keene State College in southwestern New Hampshire. She teaches a variety of geography courses including *Geospatial Technologies for Teachers,* secondary social studies methods, and supervises student teachers. Her research interests include the application of geospatial technologies in the K–12 classroom and professional development in geography education. She is a National Board Certified Teacher in Adolescent/Young Adult Social Studies-History. She can be reached at kalvarez@keene.edu.

Thomas Baker is currently the Education Industries Internet Manager at Environmental Systems Research Institute (ESRI). In his post-doctoral position at the University of Kansas, he designed the KanGIS web portal and supported international online GIS collaborative projects with the Pathfinder project. Dr. Baker co-edited the GIS-themed issue of the NCGE's *Journal of Geography* and has published many articles on GIS in education. He can be reached at tbaker@esri.com.

Digital Geography: Geospatial Technologies in the Social Studies Classroom, pages 307–314
Copyright © 2008 by Information Age Publishing
All rights of reproduction in any form reserved.

Robert S. Bednarz is Professor of Geography at Texas A&M University. He teaches both undergraduate and graduate courses and supervises graduate students who focus on his research interests—spatial thinking and the use and impact of geospatial technologies in geography education. He has published in *Professional Geographer, Journal of Geography, Journal of Geography in Higher Education, Social Education,* and *International Research in Geographic and Environmental Education.* He can be reached via r-bednarz@tamu.edu.

Sarah Witham Bednarz is a Professor in the Department of Geography at Texas A&M University. She teaches undergraduate courses on human geography, the geography of Texas, and geography education and graduate courses in theory and research in geography education. Her research interests focus on learning and teaching geography and related disciplines, particularly on the ways people learn to think spatially using different technologies. In addition, she has published on the implementation of discipline-based education reform, and the application of research to the development of effective science and social science instruction. As one of the primary authors of the National Geography Standards, she developed the materials on geographic skills as well as other components of the project. She may be reached at s-bednarz@tamu.edu

Matthew Brown is a Director of Inquirium LLC, a design group specializing in the development of tools to support learning. He received his Ph.D. from Northwestern University in the Learning Sciences and completed a postdoc at the University of Illinois at Chicago, where he also served as Assistant Director of Research at the Center for the Study of Learning, Instruction, and Teacher Development. Matt is also a former History teacher at both the high school and middle school levels. He has worked extensively in the areas of school reform, software design, and curriculum development. Matt is the author of several articles on the intersection of classroom tools and teacher practices. His design portfolio includes work for the Chicago Public Schools, Northwestern University, University of Chicago, University of Illinois, the Museum of Science and Industry, the Chicago Zoological Society, the Adler Planetarium, and Intel Education.

Rick L. Bunch is Associate Professor and Director of the Center for Geographic Information Science and Health in the Department of Geography at the University of North Carolina at Greensboro. His research and teaching interests are in Geographic Information Science and spatial cognition. He serves as Book Review Editor for *Research in Geographic Education* and as a member of the editorial board for *Cartography and Geographic Information Science.* He has published articles in the *Annals of the Association of American Geographers, The Professional Geographer, Research in Geographic Education,*

Southeastern Geographer, and Cartography and Geographic Information Science. He can be reached at rlbunch@uncg.edu.

John L. Conant is Professor and Chair in the Department of Economics at Indiana State University. Additionally, he serves as the director of the ISU Center for Economics Education.

Aaron Doering is an assistant professor in Learning Technologies at the University of Minnesota (http://education.umn.edu/CI/Faculty/Doering.html). His research involves the design, development and evaluation of online learning environments, technology integration in K–12 preservice and inservice settings, and the innovative use of technology to support teaching and learning. Most recently, his research and projects have focused on hybrid online adventure learning environments (http://www.polarhusky.com), online scaffolding environments, and online community-building environments. Aaron also researches content, cohort-specific technology integration within preservice teacher education programs. He can be reached at adoering@umn.edu.

Roger M. Downs is Professor of Geography at The Pennsylvania State University. He teaches undergraduate courses in human geography and graduate courses in research design and geography education. His research interests include the development of spatial thinking and spatial literacy, and the history of geography education. He has published articles in journals such as the *Annals of the Association of American Geographers, Cognitive Development*, and *Transactions in GIS*. He can be reached at rd7@psu.edu.

Daniel C. Edelson is Vice President for Education and Children's Programs at the National Geographic Society. Previously he founded and directed the Geographic Data in Education (GEODE) Initiative at Northwestern University. Trained as a computer and cognitive scientist, Dr. Edelson develops and studies instructional materials that are informed by contemporary research on learning and motivation. Two recent products of his integrated research and development efforts are: My World GIS, a geographic information system for inquiry-based learning, and *Investigations in Environmental Science*, a case-based, technology-infused high school environmental science text. His email address is dedelson@ngs.org.

Sven Fuhrmann is an Assistant Professor at Texas State University—San Marcos and directs the Geospatial Cognition Laboratory. Fuhrmann has over 15 years of experience in GIScience and Geovisualization, bridging cutting-edge research developments in the United States and Europe. He is an established member in national and international commissions that are concerned with developing geospatial research agendas and novel research strategies. His research interests include GIScience, 3D Geovisualization and Interaction, Human-Centered Design, Emergency Management, Spatial Cognition and Behavior.

Mark Gahegan is Professor of Geography and also Professor of Information Science and Technology at the Pennsylvania State University, USA. His undergraduate degree is in Information Systems from the University of Leeds, UK, and his Ph.D. is in the crossover area between Computer Science and Geography, *Geographical Information Science*—from Curtin University, School of Computing Science, Perth, Australia. His research interests are in computational semantics in the earth sciences, geographical information science, visualization, remote sensing, machine vision, spatial data structures and algorithms, e-science and e-education. He is the Americas editor for the International Journal of Geographical Information Science, and associate director of the GeoVISTA Center for visualization and spatial analysis research at Penn State, and is an active participant in several e-science and e-education initiatives.

Jay D. Gatrell is an Associate Professor of geography & women's studies and Associate Dean of the College of Arts & Sciences at Indiana State University. He served as social studies education coordinator between 2002 and 2006. He may be reached at jgatrell@isugw.indstate.edu.

David R. Green is Director of the Marine and Coastal Resource Management (MCRM) Degree Programme and Director of the Centre for Marine and Coastal Zone Management (CMCZM) in the Department of Geography and Environment, at the University of Aberdeen, Scotland, UK. He is a Senior Lecturer and Lectures in Geographic Information Systems (GIS), Remote Sensing, Digital Image Processing (DIP), Digital Mapping and Cartography, and Coastal Ecology, as well as environmental applications of the geospatial technologies. His research interests are wide ranging, and include coastal applications of the geospatial technologies, precision viticulture and vineyard management, journalistic cartography, and landscape ecology. He has published papers in the *Professional Geographer, Geography, Cartographica, Science of the Total Environment, Environmental Pollution, Journal of Coastal Conservation, Zeitschrift fur Geomorphologie, The Journal of Geography in Higher Education,* and *The SUC Bulletin.* He is Editor in Chief of the *Journal of Coastal Conservation: Planning and Management (Springer),* and a member of the Editorial Boards of *The International Journal of Science Education (Taylor and Francis),* and *International Journal of Earth and Environmental Sciences.* He is past Chairman of the Association for Geographic Information (AGI), a past Editor of The Cartographic Journal, and is currently President of the EUCC-The Coastal Union in Leiden, The Netherlands, and Chief Examiner of the ASET GIS qualification in the UK. His books include: *Landscape Ecology and GIS* (Taylor and Francis), *Integrated Coastal Zone Management—The Present Global Scenario (Research Publishing Services), Going Mobile: Mobile Technologies and GIS (URISA), Coastal and Marine Geo-Information Systems: Applying the Technology to the Environment (Kluwer), Coastal Remote Sensing Guidelines (URISA),* and *GIS in Secondary School Education (Taylor and Francis)*. He can be reached at: d.r.green@abdn.ac.uk

Michael J. Kane is an Associate Professor in the Department of Psychology at the University of North Carolina at Greensboro. His research and teaching interests are in cognitive psychology, generally, and in individual differences in memory and attention, specifically. He serves as Associate Editor for the journal *Memory & Cognition* , and he has published articles in such journals as *Psychological Bulletin, Psychological Science, Journal of Experimental Psychology: General* , and *Journal of Experimental Psychology: Learning, Memory, and Cognition.* He can be reached at mjkane@uncg.edu.

Joseph J. Kerski serves as Curriculum Development Manager for Environmental Systems Research Institute (ESRI) Inc. He creates GIS-based curricular materials at all levels, teaches online and face-to-face geotechnology courses and workshops nationally and internationally for university, secondary, and primary faculty, and for students, scientists, and public and private practitioners, and partners with professional organizations to advance GIS throughout education and society. His research interests include the effectiveness and implementation of geotechnologies in education. With three degrees in geography, Joseph has served in three federal agencies and taught at the University of Denver and at Sinte Gleska University on the Rosebud Sioux reservation. He is co-author of the book *Essentials of the Environment* and can be reached at jkerski@esri.com.

Robert Earl Lloyd is Distinguished Professor Emeritus in the Department of Geography at the University of South Carolina. His research and teaching interests are in spatial cognition, quantitative methods, and cartography. He has published articles in geography journals such as *Annals of the Association of American Geographers, The Professional Geographer, Research in Geographic Education,* and *Economic Geography* and in GIS journals such as *Cartography and Geographic Information Science, Cartographica, Cartographic Perspectives, The Cartographic Journal* and *The International Journal of Geographical Information Science.* He can be reached at blloyd@sc.edu.

Alan M. MacEachren is 2004–2007 E. Willard and Ruby S. Miller Professor of Geography and Director of the GeoVISTA Center (www.GeoVISTA.psu.edu) at the Pennsylvania State University. His research foci include: visual analytics, geovisualization, geocollaboration, interfaces to geospatial information technologies, spatial cognition, human-centered systems, and user-centered design. He served as chair of the International Cartographic Association *Commission on Visualization and Virtual Environments* (1999–2005) and was named honorary fellow of that organization in 2005. He was a member of the National Research Council, Computer Science and Telecommunications Board *Committee on the Intersections Between Geospatial Information and Information Technology* (2001–2002) and of the National Visualization and Analytics Center *R&D Agenda panel* (2004–2005). MacEachren is author of *How Maps Work: Representation, Visualization and Design*, Guilford Press, 1995 and *Some Truth with Maps*, Association of American Geographers, 1994 and is co-editor of additional books (including *Exploring Geovisualization*, Elsevier, 2005).

Andrew J. Milson is an Associate Professor in the Department of Teacher Education and Administration at the University of North Texas. He teaches undergraduate courses in secondary education and graduate courses in curriculum studies. His research interests include the implementation of inquiry learning methodologies in secondary schools and the integration of the Internet and geospatial technologies into social studies education. He served as co-editor of the volume *Readings in American Educational Thought* (Information Age, 2004) and has published articles in journals such as *Theory and Research in Social Education*, *Journal of Educational Research*, *Journal of Geography*, *Educational Policy*, *Social Education*, and *The Social Studies*. He can be reached at andy.milson@unt.edu.

Joanna Mouatt is a geologist for Marathon International Petroleum (G.B.) Limited, based in Aberdeen, Scotland. She has a Ph.D. in geology from the University of Manchester (U.K.) specializing in the geochemistry of igneous rocks. An interest in internet GIS and digital globe applications, such as Google Earth, as a tool for teaching and distributing geospatial information was developed while completing a Postgraduate Diploma in GIS at the University of Aberdeen, UK. She can be contacted at joanna.mouatt@gmail.com.

Elisabeth S. Nelson is an Associate Professor in the Department of Geography at the University of North Carolina at Greensboro. Her research and teaching interests are in map design and map perception and cognition. She has published in *Cartography and Geographic Information Science*, *Cartographic Perspectives*, and *Cartographica*. She currently serves as the editor of *Cartography and Geographic Information Science*. She may be reached at esnelso2@uncg.edu.

Josh Radinsky is an Assistant Professor of the Learning Sciences at the University of Illinois at Chicago. He teaches courses in research on learning environments and social studies education, and conducts research in collaboration with teachers in Chicago public schools. His research uses sociocultural and cognitive theoretical frameworks to study how students learn to reason with visual data, with an emphasis on history and the social sciences in elementary, secondary, and undergraduate settings. His research and development projects include *Teaching and Learning with GIS* (http://litd.psch.uic.edu/gis/) and *GIS for History* (http://www.gisforhistory.org). He can be reached at joshuar@uic.edu.

Jennifer A. Roberts is a Ph.D. candidate in the Curriculum & Instruction program of the Department of Teacher Education and Administration at the University of North Texas. Her research interests include teacher quality and preparation, professional development, and adult learning. Prior to becoming a full-time doctoral student, she taught elementary school in the north Texas area. She can be reached at jroberts@coe.unt.edu.

Cassandra Scharber is a doctoral student in the Department of Curriculum and Instruction at the University of Minnesota. Majoring in both learning technologies and literacy, her research interests include K–12 technology integration, digital equity, digital literacies, and automated essay scoring. A former secondary English teacher, she has expertise in developing online curricula for both K–12 students and preservice teachers. She can be contacted at scha0809@umn.edu.

Eui-kyung Shin is an Assistant Professor of Social Studies Education in the Department of Teaching and Learning at Northern Illinois University. She teaches undergraduate courses in elementary education and graduate courses in global education. Her research interests include the integration of Geographic Information System (GIS) into elementary classrooms and the infusion of global perspectives into the elementary curriculum. She has published articles in journals such as *Journal of Geography, Theory and Research in Social Education, Action in Teacher Education, The Social Studies,* and *Social Studies and the Young Learner.* She can be reached at ekshin@niu.edu.

David A. Smith is currently an Educational Designer for the GLOBE Program acting as the project lead on the new Watershed Dynamics Earth System Science project. His responsibilities include the design and development of a multiyear curriculum using geospatial technologies in the science classroom. Prior to GLOBE, he was a curriculum developer at Northwestern University with primary responsibilities in the development of curriculum modules and the support and implementation of the My World GIS software program developed under the GEODE Initiative at Northwestern. Prior to work at NU, he worked with students and teachers participating in the Orton Family Foundation's Community Mapping Program in Colorado and developed education and outreach activities for the

Digital Library for Earth System Education (DLESE) Program Center in Boulder Colorado.

Thomas A. Tricot is a Ph.D. candidate in the Department of Geography at the University of North Carolina at Greensboro. His research and teaching interests are in Geographic Information Science and Geographic Education. He can be reached at tatricot@uncg.edu

George Veletsianos is a Ph.D. student in Learning Technologies at the University of Minnesota. His research interests revolve around the design, development, and evaluation of electronic learning environments. Specifically, he is interested in pedagogy, social media, and how learners interact, collaborate, and learn with virtual characters. As such, George's research is cross- and multi-disciplinary encompassing ideas from learning technologies, social psychology, human-computer interaction, communication sciences, and computer science. He can be reached at velet006@umn.edu.

Shannon White is Assistant Director of the Florida Center for Instructional Technology (FCIT) at the University of South Florida in Tampa. Dr. White received her Ph.D. at North Carolina State University in Instructional Technology. Her dissertation was entitled, "Geographic Information Systems (GIS) and Instructional Technology (IT) Diffusion: K–12 student and educator conceptualizations." While at NCSU, she taught undergraduate and graduate geography, GIS in Education and Social Studies Methods courses. Dr. White directs the Florida Digital Educator program, a statewide technology integration and professional development initiative. She may be reached at swhite@coedu.usf.edu.

9 781593 116729

Printed and bound by CPI Group (UK) Ltd, Croydon, CR0 4YY

15/09/2026

14962367-0002